Computational Science

An Introduction for Scientists and Engineers

Christopher D. Wentworth
Professor Emeritus of Physics
Doane Universisty

2025, First Edition

ISBN 979-8-9928002-1-0 (pbk)

Publisher's Cataloging-in-Publication Data

Title: Computational science: an introduction for scientists and engineers /
Christopher D. Wentworth.
Description: First edition. | Crete NE : C.D. Wentworth, 2025. | Includes
approximately 100 color photos, charts, graphs, and diagrams. | Includes
bibliographic references and index. | Summary: An introduction to
computational problem-solving, mathematical modeling, simulation, and
scientific visualization for students in the natural and engineering sciences
using the Python programming language. No programming experience is
assumed.
Identifiers: LCCN 2025908202 | ISBN 9798992800210 (pbk) | ISBN 9798992800203
(pdf) | ISBN 9798992800227 (epub)
Subjects: LCSH: Computer science – Textbooks. | Computer programming –
Textbooks. | Computer simulation – Textbooks. | Electronic data processing –
Textbooks. | Mathematical models – Textbooks. | Python (Computer program
language). | BISAC: COMPUTERS / Computer Science. | COMPUTERS /
Computer Simulation. | MATHEMATICS / Applied. LC record available at
https://lccn.loc.gov/2025908202

Book Website: ComputationalScienceBook.info

Acknowledgements

In the many years I have taught computational science there are many people to whom I am indebted. I will list a few here.

Most important are the many students who experienced me as their instructor. They asked the critical questions and provided valuable feedback about explanations and assignments that proved helpful to them. I have been privileged to hear back from several of these students as they progressed in their careers letting me know about the challenges and successes in applying computational science concepts in their work. All of these communications with students were central to how I wrote this book.

Colleagues at Doane University who helped develop an interest in computational science and offered significant advice on how it should be structured include Tessa Durham Brooks, Susan Enders, Alec Engebretson, Peg Hart, and Mark Meysenburg. Lavi Zamstein proofread several of the chapters and caught mistakes. Remaining errors are, of course, my responsibility.

I greatly benefited from resources provided by The Shodor Education Foundation, in particular, conversations with Robert Panoff, President and Executive Director of the foundation.

Finally, I want to thank my wife, Sally, for her patience and encouragement throughout the writing process, particularly while I was being treated for leukemia.

Contents

Preface

Computational science involves using mathematics, computer science, and domain knowledge to solve complex problems in the domain of interest. The ability to represent real-world systems with mathematical models and use computers to simulate and analyze the behavior of these models is now an essential part of research in the natural, social, and engineering sciences and in design work performed by engineers. The goal of this book is to introduce new students in science and engineering to fundamental concepts and tools of computational science and apply these ideas to example problems in several areas of science and engineering.

We will begin our study of computational science with an introduction to computational thinking, which is concerned with mental skills and practices that facilitate using computation to solve problems. Computational thinking is then applied to developing a strategy to create mathematical models that can be studied using computer simulation. Next, basic programming concepts are introduced within the context of a science or engineering problem. We introduce the Numpy package, which is critical for performing numerical computations in Python. Principles of scientific visualization are developed with examples created using Python visualization packages. We introduce the concept of a dynamical system and how to develop a model of such a system. We develop techniques for the numerical simulation of dynamical system models. Next, we investigate several kinds of stochastic models and using Monte Carlo methods to study such models. The book concludes with several project suggestions that require practicing the modeling and programming tools developed in the rest of the book.

Many books on computational science or computational methods in science and engineering assume that the student has a well-developed mathematical background, including courses in calculus and differential equations. Since the goal of this book is to expose first-year students to computational science techniques, and these students have typically not taken calculus in high school, we assume the student is ready to take calculus but has not yet done so. Some calculus concepts are introduced in the book, but mainly through geometrical interpretation rather than through rigorous mathematical definitions. Dynamical systems models, defined using differential equations, are introduced, but their properties are explored primarily through computer simulation rather than through analytical techniques.

No programming experience is assumed by the book. The necessary techniques are developed simultaneously with computational science ideas. Students will gain programming experience in the context of using programming to solve science and engineering problems.

Two kinds of programming exercises are provided at the end of each chapter: program modification problems and program development problems. The modification problems require adapting a code developed in the chapter to a new but related problem. The development problems ask students to solve a non-trivial problem not yet addressed in the text, including developing all code required for the solution. These development problems will often require integrating ideas from previous chapters in addition to ideas developed in the chapter posing the problem.

Students who master the concepts developed in this book and complete the program modification and program development problems should be ready to use computation in many advanced courses required by science and engineering majors. Students should also be ready to increase their knowledge of computational science through the independent study of many library and online resources.

1. Introduction

1.1. Defining Computational Science

Before embarking on our journey to learning how computational science is used to solve scientific and engineering problems, we should first establish more formally what we mean by computational science. When you think of the word "computation," what comes to mind? Perhaps you recall practicing long division in elementary school, balancing your checkbook, or estimating groceries required for a dinner party. These are instances of equating computation to arithmetic, but the modern conception of computation in science and engineering goes well beyond this concept.

We will begin our overview of computational science by defining it and showing the breadth of disciplines and enterprises that make use of it to solve problems. Computational science sounds a lot like computer science, but it is a different discipline, although it does involve computer science. Here is a good working definition that we will use:

__Computational science__ is the development or application of techniques that use computation to solve scientific and engineering problems.

There are certainly other definitions that can be found in the literature, but this definition will work for us.

Computational science is inherently interdisciplinary, involving concepts and techniques from computer science, mathematics, and a domain-specific science or engineering discipline.

Traditionally, a scientific discipline involves two reinforcing ways of working: experiment (or observation) and theory. Figure 1.1 illustrates these two essential aspects of science.

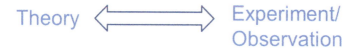

Figure 1.1.: Traditional concept of science.

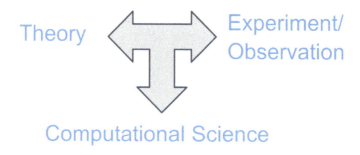

Figure 1.2.: The modern concept of science.

The existence of powerful computing technologies has changed the conduct of scientific and engineering work profoundly by allowing large data sets to be studied and realistic models of systems to be simulated. Computation aids both theoretical and experimental work, so that computational science becomes an indispensable third leg of our concept of scientific work, as illustrated in Figure 1.2.

Consider the scientific discipline astronomy. Current observational techniques have created huge amounts of data that include objects located (stars, galaxies, exoplanets, etc.), motion data of each object, intensity of electromagnetic radiation at different wavelengths associated with these objects, and other properties. Table 1 shows the amount of data generated by several observational projects. Note that TB stands for terabyte (10^{12}), PB stands for petabyte (10^{15}), and EB stands for exabyte (10^{18}). We see that each of these individual projects creates enormous amounts of data.

Cataloging, storing, uncovering systematic errors, and performing exploratory data analysis all require significant computational resources and specialized techniques.

Table 1.1.: Data volumes of different sky survey projects (Zhang & Zhao, 2015).

Sky Survey Projects	Data Volume
DPOSS (The Palomar Digital Sky Survey)	3 TB
2MASS (The Two Micron All-Sky Survey)	10 TB
GBT (Green Bank Telescope)	20 PB
GALEX (The Galaxy Evolution Explorer)	30 TB
SDSS (The Sloan Digital Sky Survey)	40 TB
SkyMapper Southern Sky Survey	500 TB
LSST (The Large Synoptic Survey Telescope)	*~200 PB expected*
SKA (The Square Kilometer Array)	*~4.6 EB expected*

Another way to think of computational science is that it involves the intersection between computer science, mathematics, and an applied discipline, as illustrated in Figure 1.3.

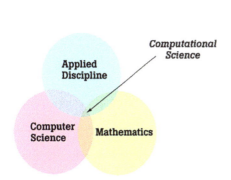

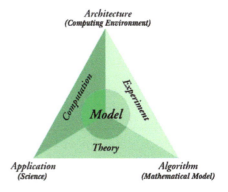

Figure 1.3.: Conceptual model of computational science (The Shodor Education Foundation, 2000).

Figure 1.4.: The relationships between models, the basic elements of science, and elements of computational science (The Shodor Education Foundation, 2000).

Mathematical modeling is a central activity for working physicists and engineers. As other sciences have become more quantitative, mathematical modeling has assumed an increasingly important role in those disciplines as well. This includes areas such as biology and environmental sciences. A significant amount of work in contemporary computational science involves developing and exploring mathematical models using computation. This focus will be important to us throughout this course, so it is worth keeping in mind how modeling connects with other aspects of doing science. Figure 4 illustrates the connections.

A model starts with experimental data or observations and describes patterns in the data, connecting those patterns with more fundamental theoretical ideas from the area of science being investigated. The process of uncovering patterns typically involves elements of computation such as statistical analysis and visualizations. The outer elements of Figure 1.4 illustrate key elements of computational science itself: a scientific activity (application) that is facilitated by using mathematics and algorithmic thinking (algorithms), which, in turn, requires computation to fully explore (computing environment).

1.2. Examples of Computational Science Applications

Let's look at some specific examples of scientific and engineering problems that use computation science.

1.2.1. Global Climate Change

Climate change is an important topic that impacts many areas of everyday life and promises to have even more profound effects in the future. It is important for us to understand how to describe climate, recognize the important determinants of climate, and develop realistic models that can help us predict the future climate.

Describing the global climate starts with some basic observational data such as the average surface temperature. Figure 1.5 shows data on the average surface temperature. Notice that the graph does not directly show the temperature, but rather the temperature anomaly, which is the difference between the actual temperature and the average temperature measured over a specified base time period. Using the temperature anomaly avoids including effects of systematic errors in an individual thermometer.

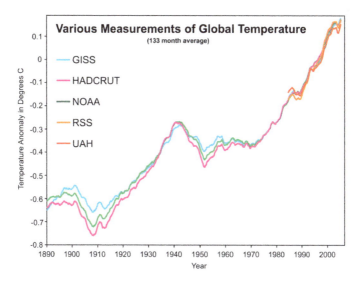

Figure 1.5.: Average surface temperature anomaly (*Temperature Composite*, n.d.).

Producing just this one graph involved many aspects of computational science.

■ The individual data sets contain a large number of measurements that must be made available in a user-friendly form.

■ Each data set must undergo statistical analysis to uncover and correct for problems in the data.

■ A moving average is calculated to remove some of the random variation present in the observations. Such smoothing of the data helps to identify trends

■ The graph of the moving average must be generated.

Each of these actions requires computation to accomplish.

The graph helps us to identify patterns, namely, that average global surface temperature increased from 1910 to 1940, leveled off and decreased slightly between 1940 and 1950, then increased again at a faster rate from 1950 to the present.

Another way of visualizing the data that requires significant computational work is to show the temperature anomaly in different regions of the earth and follow the geographic distribution over time. Figure 1.6 shows the result of performing this kind of analysis. The level of the temperature anomaly is indicated by the intensity of the color.

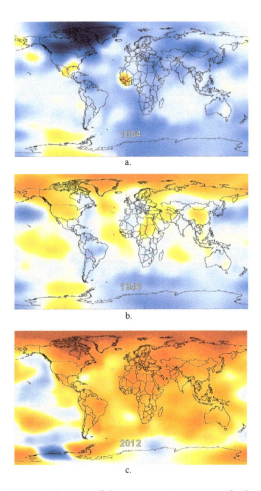

Figure 1.6.: Geographic distribution of the temperature anomaly (NASA's Scientific Visualization Studio, 2021).

Creating visualizations of the temperature data is only the beginning of investigating climate change. Models based on fundamental science including physics, chemistry, and meteorology must be constructed to help us understand the underlying causes for the changes we see. These models typically are mathematically sophisticated and can usually only be explored using numerical calculations performed by a computer.

1.2.2. Evolution of Cosmic Structures

Understanding how specific cosmic structures, such as galaxies, evolve over the history of the universe is an important problem of current interest. Observational astronomers can locate and characterize galaxies that have existed since a few million years after the Big Bang. Figure 1.7 shows some of the different kinds of galaxies found in the universe. Progress in understanding how the observed distribution of galaxy morphologies has recently been made by using computer simulations based on fundamental astrophysics principles incorporating hydrodynamics and gravitation principles (Vogelsberger et al., 2014). The researchers began with a system approximating the state of the universe at 12 million years after the Big Bang.

Figure 1.7.: Three general galaxy morphologies with specific examples of each. (Feild, 2019).

The computer simulation over about 13 billion years of evolution was able to show realistic distributions of galaxy morphologies. Figure 1.8 shows some of the simulated galaxies produced from the simulation.

1.2.3. Designing the Boeing 777

The Boeing 777, shown in Figure 1.9, is a twin-engine, wide-body, commercial jet that was first introduced in 1994 but is still in production.

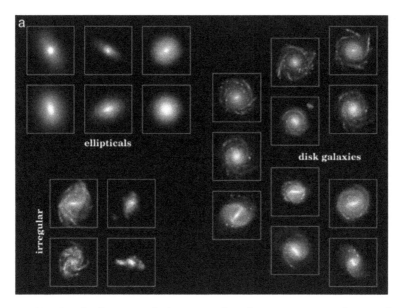

Figure 1.8.: Simulated galaxy morphologies produce by computer simulation of galactic structure evolution (Vogelsberger et al., 2014).

It was a significant achievement of engineering because it was the first airplane to be completely designed using computer-assisted design tools (CAD). This eliminated the need for most full-scale mock-ups. This reduced the time and cost of designing and revising all the systems involved in the aircraft. The CAD system combined significant engineering mechanics knowledge with extremely sophisticated computer graphics to ensure that all parts of the system fit to within required tolerances and could move correctly. The CAD system also had properties of the materials used for parts built in so that stresses and strains could be measured in the simulated system to ensure part failure would not occur. Each system of the plane could be simulated through multiple cycles of revisions with both software and human-based visual checks. Figure 1.10 shows examples of the visual product from the CAD system.

Every major area of science and engineering now has an established subarea that is focused on developing appropriate computational approaches to problem-solving in the particular discipline. The following is a list of such disciplines:

Figure 1.9.: . The Boeing 777 commercial jet (Koske, 2009).

Figure 1.10.: Graphics output from the CAD software used for designing the Boeing 777 (Dietrich et al., 2005).

- Computational Biology
- Computational Chemistry
- Computational Economics
- Computational Finance
- Computational Neuroscience
- Computational Physics
- Computational Sociology

1.3. Elements of Computational Thinking

While the term "computational thinking" is relatively new, having been popularized in 2006 by Jeannette Wing in an opinion piece about computer science education (Wing, 2006), but concepts associated with the term have a long history. The ancient Babylonians (1800-1600 BCE) developed tables of reciprocals and multiplications to speed up and systematize arithmetical calculations. They also developed step-by-step procedures (i.e., algorithms) for solving certain classes of algebra problems (Knuth, 1972). Figure 1.11 shows an example of a Babylonian clay tablet specifying the procedure for solving a mathematical problem (Amin, 2019).

1.3.1. Historical Examples of Computational Thinking

The Ancient Greek astronomer Hipparchus (190-120 BCE), shown in Figure 1.12, developed a theoretical framework to predict positions of sun and moon as functions of time. This framework used relationships from spherical trigonometry, and the required calculations used regular trigonometric functions such as cosine and sine. To expedite calculations involving trig functions, Hipparchus developed procedures for producing trigonometric function tables. Such tables remained useful well into the 20th century before being replaced by electronic calculators.

An example of a spherical triangle is shown in Figure 1.13.

The Persian mathematician Muhammad ibn Musa al-Khwarizmi, shown in Figure 1.14, wrote an important book published around 800 CE that specified procedures for solving quadratic equations. The Latin translation of his book on the Hindu-Arabic numeral system, *Algoritmi de numero Indorum*, gave us the term "algorithm", which will play a significant role in our overview of computational thinking.

There is not a universal definition for the term "computational thinking", yet it is really central to your development as a computational scientist. So, how can you be expected to learn how to do something so important when it cannot be defined? Life is truly unfair. But do not despair; we will look at some elements of computational thinking that most computational scientists will agree are important elements of learning how to approach problems so that they can be solved

Figure 1.11.: Babylonian clay tablet describing a mathematical procedure (Amin, 2019).

Figure 1.12.: The Greek astronomer Hipparchus (New York Public Library / Science Source / Science Photo Library, n.d.).

computationally. Indeed this can be a starting point for our working definition of computational thinking. Here is how Jeannette Wing of Columbia University defines it:

> Computational thinking is the thought processes involved in formulating a problem and expressing its solution(s) in such a way that a computer—human or machine—can effectively carry it out (Wing, 2014).

An athlete must train their body to execute physical motions required by the sport. A computational scientist must train their mind to think about problems in a way that allows computers to help solve the problems. Computational thinking is not used only for solving problems with a computer but can provide strategies for solving all sorts of problems. The habits of mind and processes you will learn here can be widely applied.

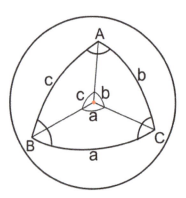

Figure 1.13.: Example of triangle on a sphere (Mercator, 2013).

Figure 1.14.: Commemorative Stamp of Muḥammad ibn Musā al-Ḵwārizmī (unknown, 1983).

1.3.2. Uses of Computational Thinking

Computational thinking is used in many disciplines and activities with significant real-world applications:

- Mapping the human genome
- Predicting the spread of infectious disease
- Developing methods for weather prediction
- Predicting the effect of government policies
- Developing a method of scheduling meetings efficiently in a large organization

1.3.3. Pillars of Computational Thinking

The general principles of computational thinking can be represented using four pillars, or principle concepts. They are

- Decomposition
- Pattern Recognition
- Data Representation and Abstraction (we can shorten this to just Abstraction)
- Algorithms

Let's consider each of these pillars.

Decomposition involves breaking a complex problem into smaller, more-manageable sub-problems or parts. Consider the problem of writing a paper for a course. We might break this problem into at least three parts: writing an introduction, writing the main body, and finally writing a conclusion.

Activity: Consider the problem of cleaning your apartment or house. Apply the process of decomposition to this problem.

Pattern recognition involves finding similarities or shared characteristics within or between problems. If two problems share characteristics, then the same solution can be used to solve both problems.

Activity: Think of a problem for which you could use computational thinking, describe it, and then describe how you would apply pattern recognition.

The third pillar of computational thinking is called **data representation and abstraction**. We will shorten this to just **abstraction**. The key element of abstraction is to determine the characteristics of a problem that are important and those that can be ignored. Next, we decide on a way to represent the chosen important characteristics. This would be the data representation.

Example: developing an online book catalog. Important information:

- Title
- Authors
- ISBN
- Year of publication
- Edition

Unimportant information:

- Book jacket color
- Author birthplace

Activity: Think of a problem for which you could use computational thinking, describe it, and then describe how you would apply data representation and abstraction.

The final pillar of computational thinking is **algorithms**. We will explore algorithms throughout this course. For now, we define their essential properties as:

1. A collection of individual steps to be used in solving the problem;
2. Each step must be defined precisely;
3. The steps must be performed sequentially.

We can represent an algorithm using a flow chart.

The elements of flowcharts include

1. Terminals: an oval that represents a start or stop.

2. Input/Output: a parallelogram that represents the program either getting data input or creating data output.

3. Processing: a box that represents executing some action such as performing arithmetic.

4. Decision: a diamond represents making a decision.

5. Flow Lines: arrows that indicate the sequence taken by instruction. The arrows are called directed edges by computer scientists.

Another way of representing an algorithm is using pseudocode, which we will explore later.

1.3.4. Logical Arguments

A good starting point for computational thinking is the idea of logical thinking. Logic is a systematic approach to distinguish between incorrect and correct arguments. Logical reasoning starts with premises, which are statements about initial knowns concerning the problem. Premises must be determined to be true or false. If the premises are true then the final step in a logical argument is to reason that a conclusion follows.

There are two general categories of logical arguments with which we are concerned: deductive **and** inductive **arguments**. In a deductive argument, once the premises are established to be true then the conclusion will necessarily follow with complete certainty. In an inductive argument the veracity of statements, whether they be premises or a conclusion, can be expressed only with a particular probability rather than with certainty.

Example of deductive reasoning:

1. Socrates is a man.
2. All men are mortal.
3. Therefore, Socrates is mortal

Statements 1 and 2 are the premises. If they are both true then the conclusion necessarily follows.

A deductive argument can fail in two ways:

1. One or more of the premises could be false, which means that the conclusion is not necessarily true.

6. The conclusion does not necessarily follow from the premises, even if they are all true. There is a logic problem in the argument.

Example of inductive reasoning:

In an inductive argument there will be an element of probability involved in the truth of premises and conclusion. In an inductive argument we establish that the conclusion is probably true but we cannot say it is true with certainty, as we can with a deductive argument. Here is an example:

90% of the senior class at Belmont High were accepted into college.

Robert is a senior at Belmont High

Therefore, Robert was accepted into college.

The conclusion is probably true but not necessarily true.

Computers most easily perform deductive reasoning. In the next Module, we will learn about a particular way of performing deductive arguments called Boolean logic that is implemented in all programming languages, including Python.

1.4. Mathematical Modeling in Science and Engineering

Modeling is a fundamental element of science and engineering. We can think of a model as a simplified representation of a system of interest. A model could be physical such as letting a basketball represent the sun and a golf ball represent the earth and carry the golf ball around the basketball to illustrate a planetary orbit. Automobile manufacturers historically used clay models of a vehicle to explore how the shape influenced airflow around the vehicle. In science and engineering models are usually represented using mathematics such as representing the position of a planet by its spatial coordinates (r, θ, ϕ) in a spherical coordinate system and then representing the orbit by the equation for an ellipse, Equation 1.1.

$$r = \frac{a\left(1 - e^2\right)}{1 + e \cos\left(\phi\right)} \tag{1.1}$$

We can usefully classify the kinds of models used in science and engineering into the following categories.

- Deterministic
- Stochastic
- Empirical
- Theory-based

1.4.1. Deterministic Models

Deterministic models describe systems in which knowledge of the state at one time will determine the state at subsequent times. There is no element of randomness in the state of the system. An example would be applying Newton's laws of motion to a rock thrown in the air. Knowing the rock's position and velocity at one time will allow us to predict the position and velocity at subsequent times. Figure 1.15 illustrates this situation.

1.4.2. Stochastic Models

Stochastic models involve some element of randomness. When such models are described mathematically, they will involve use of probability theory. An example is modeling the trajectory of smoke coming from a burning candle, as shown in Figure 1.16. Candle smoke (Wicks, 2022). The first part of the smoke trajectory after it leaves the wick could be described using a deterministic model, but most of the smoke motion involves an element of randomness, suggesting that a stochastic model would be required.

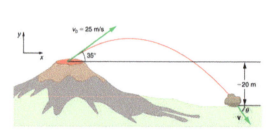

Figure 1.15.: Trajectory of a rock ejected from a volcano. (Urone & Hinrichs, 2012)

Figure 1.16.: Candle smoke (Wicks, 2022).

1.4.3. Empirical Models

Empirical models attempt to find the mathematical form relating variables used to describe raw observational data or experimental results. The model is purely descriptive and not interpreted as an explanation of a pattern based on more fundamental principles. For example, suppose we take stroboscopic photographs of a falling ball at equal time intervals, as shown in the photo of Figure 1.17. If we measure the ball's position as a function of time from the photograph and make a graph, we will obtain the graph in Figure 1.17.

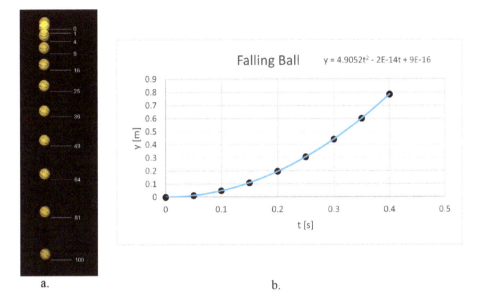

Figure 1.17.: (a) Stroboscopic photographs of a falling ball (Maggs, 2007). (b) Graph of position data.

Looking at the graph of the position data taken from strobe photos, we would rule out a linear relationship between y and t. The next type of equation to try would probably be a second-order polynomial, a quadratic. Such a model does fit the data well, as shown by the solid line in the graph.

1.4.4. Theory-based Models

Theory-based models use the more fundamental elements of a scientific theory, such as Newton's Laws of Motion in physics, to develop the mathematical model of a system. In the example of the dropped ball, described above, we can apply Newton's Second Law and Newton's Law of Gravitation to the ball and predict that that relationship between the ball's position and time should be a second-order polynomial equation, assuming that we ignore the effects of air resistance. We would arrive at the same equation as we hypothesized just on the basis of looking at the data, but the equation now is seen to be implied by the theory called Newton's Laws of Motion.

The categories described above are not exclusive of each other. We can have theory-based deterministic models or theory-based stochastic models, for example.

We will use a variation on our four-step problem-solving strategy discussed in Chapter 2 when developing a mathematical model of a system. Here are the basic steps.

1. Analyze the problem
2. Formulate a model
3. Solve the model
4. Verify and interpret the model's solution
5. Report on the model
6. Maintain the model

Our discussion of dynamical systems models will illustrate the use of this strategy.

1.5. Introduction to Textbook Computing Environment

Many principles of computational science are independent of the programming language used to implement the computations, but it is easier to learn these principles using specific examples that involve a particular programming language. We will use Python as our primary programming language. To enable quick development and application of computational science principles we should establish a working Python programming environment now, so that it can be used immediately.

1.5.1. Types of Programming Environments

There are several ways to establish a good programming environment that uses Python:

1. We can download and install a Python distribution, such as the Anaconda Distribution and then use the IPython command line shell that comes with that distribution. This allows the user to execute Python interactively, one code line at a time.Install a Python distribution, such as Anaconda, that comes with a Jupyter server then use Jupyter notebooks to compose exe-

cutable code and text blocks.

2. Install a Python distribution, such as Anaconda, then use a plain text editor (notepad, gedit, TextEdit) application to compose code and then execute the code from the command line using the appropriate command line console for your operating system.

3. Install a Python distribution, then install an IDE application (PyCharm, Spyder, VSCode) that bundle a smart text editor, an interactive console, and enhanced debugging tools into a nice GUI.

4. Use a cloud computing platform such as CoCalc, pythonanywhere, or Google Colab that allows the user to use Jupyter notebooks without installing any software, except for a browser. The cloud service maintains the Python distribution.

1.5.2. The Google Colab Environment

The last method is easiest and quickest to use, so we will focus on it here. The other methods are discussed in the appendix. We will use Google Colab for our example. To use Jupyter notebooks in Colab do the following.

1. Sign into your Google account.

2. Go to your Google Drive and create a folder that will contain your Colab notebooks, which will be named Colab Notebooks. Open the folder that you just created.

3. Open a new browser tab and goto https://colab.research.google.com. You will see the following window showing available file spaces:

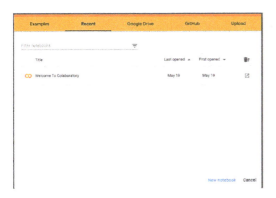

Figure 1.18.: Initial Google Colab window.

4. Select the examples tab and select the Overview of Colaboratory Features notebook.
5. Read through this notebook to get oriented to the Colab Jupyter notebook features.

Note that there are two kinds of cells in the notebook: text cells and code cells. Code cells contain executable Python code. The cell contents can be executed by selecting the Play icon on the left of the cell or by clicking anywhere in the cell and typing **Shift+Enter**.

Let's now create a new Colab notebook from your Google Drive folder.

Navigate to your Colab notebook folder that you created previously:

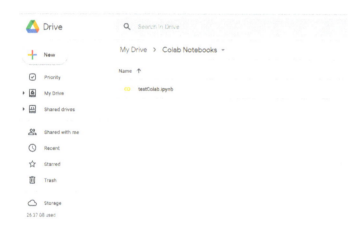

Figure 1.19.

Select

+ New

More

Google Colaboratory

Figure 1.20.

You will get a new Colab Jupyter notebook in your folder, as shown in Figure 1.21.

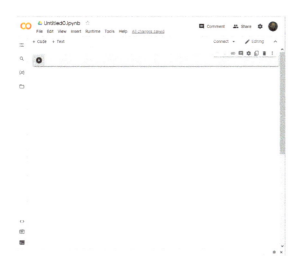

Figure 1.21.: New Colab notebook.

The new notebook is named Untitled0.ipynb by default. You can change the name as you would any Google Doc file. Leave the ipynb extension since that indicates that the file is a Jupyter notebook file.

The new notebook starts with one code cell. It is good practice to include documentation in your notebook, so let us put a text cell above the first code cell.

Place the mouse pointer right above the code cell so that the add code or text buttons appear.

Choose the **+ Text** option.

Figure 1.22.

A text cell will appear above the code cell. Double click in the cell to begin text editing.

You should see an editing box on the left and a preview of the rendered text on the right.

Figure 1.23.

You can provide formatting of the text using Markdown syntax. The example below shows how to create a heading. We will discuss Markdown syntax later.

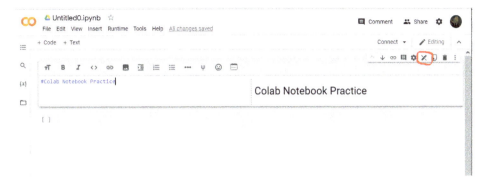

Figure 1.24.

After you finish entering any text then click the Close Markdown icon (circled above).

When you create a Colab notebook you can easily save the file in your Google Drive by selecting

File > Save

Colab will also create a temporary folder accessible to the notebook while it is open. You can view the contents of this folder by selecting the File Explorer tool on the left.

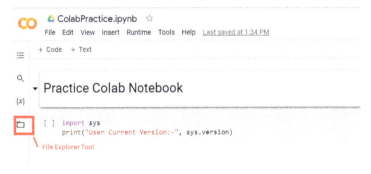

Figure 1.25.

This folder will disappear when the notebook file is closed, so if your code writes to a file you will want to ensure that the file gets placed where it will not be lost. One way to do this is to connect your Google Drive to the Colab notebook while it is running. This can be done using the Mount Google Drive tool.

Figure 1.26.

You will now see your Google Drive as a folder in the File Explorer area.

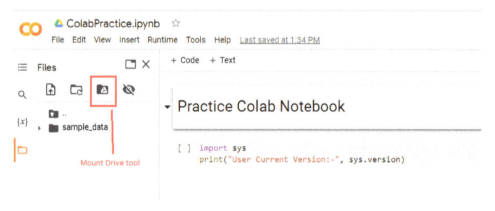

Figure 1.27.

Later, we will learn how to write directly to a file in the MyDrive folder using Python code in the notebook.

1.6. Exercises

1. Choose an area of science or engineering that interests you. Table 1 below lists areas that have a well-established record of computational solutions. Perform a web-based search using Google and library search resources to find a specific problem in the area that required computation to solve. Create a bibliography of these sources.

Table 1.2.: Wikipedia pages for areas of computational science.

Computational archaeology	Computational mathematics
Computational astrophysics	Computational mechanics
Computational biology	Computational neuroscience
Computational chemistry	Computational particle physics
Computational materials science	Computational physics
Computational economics	Computational sociology
Computational electromagnetics	Computational statistics
Computational engineering	Computational sustainability
Computational finance	Computer algebra
Computational fluid dynamics	Financial modeling
Computational forensics	Geographic information system (GIS)
Computational geophysics	High-performance computing
Computational history	Machine learning
Computational informatics	Network analysis
Computational intelligence	Neuroinformatics
Computational law	Numerical weather prediction
Computational linguistics	

You should create a document that includes the following:

■ A paragraph that describes the problem.

■ A paragraph that describes why computation was required for solving or investigating the problem.

■ Four (or more) references that give information on the problem. At least two of the references should be peer-reviewed sources such as research journal papers and academic books.

2. Computational science can be thought of as the intersection between a scientific or engineering discipline such as biology, mathematics, and _____-_____ .

3. Computer science is the study of algorithms whereas computational science is the study of how _____ can be used to solve scientific problems.

4. Traditional science would study a system by experimenting with the actual system and developing theories to help understand the system. Computational science adds to this approach by using _____ to analyze and visualize data and to explore properties of mathematical _____.

5. Which of the following are considered pillars of Computational Thinking, as defined in this course? (Choose all that are true.)

    a algorithms
    b arithmatic
    c Python
    d abstraction
    e decomposition
    f applied science
    g pattern recognition

6. Algorithms can be represented using

    a flowcharts
    b graphs
    c equations
    d mathematical variables

7. Select the statements below that are true concerning deductive reasoning.

a The conclusion is probably rather than certain.
b The conclusion necessarily follows from the premises.
c An argument could be valid but unsound.

1.7. References

Amin, O. S. M. (2019). *Babylonian Clay Tablet* [Photograph]. Own work.
https://commons.wikimedia.org/wiki/File:Clay_tablet,_mathematical,_geometric-
algebraic,_similar_to_the_Pythagorean_theorem._From_Tell_al-Dhabba%27i,_-
Iraq._2003-1595_BCE._Iraq_Museum.jpg

Dietrich, A., Wald, I., & Slusallek, P. (2005). Large-scale CAD Model Visualization
on a Scalable Shared-memory Architecture. In G. Greiner, J. Hornegger, H.
Niemann, & M. Stamminger (Eds.), *Vision, Modeling, and Visualization 2005* (pp.
303–310). Akademische Verlagsgesellschaft Aka.

Feild, A. (2019). *Galaxy Types.*
https://hubblesite.org/contents/media/images/4508-Image.html

Knuth, D. E. (1972). Ancient Babylonian algorithms. *Communications of the ACM,
15*(7), 671–677. https://doi.org/10.1145/361454.361514

Koske, K. (2009). *United Airlines B777-222 N780UA.* United Airlines Boeing
777-222 (N780UA) Uploaded by Altair78.
https://commons.wikimedia.org/wiki/File:United_Airlines_B777-222_-
N780UA.jpg

Maggs, M. (2007). *Falling Ball* [Digital].
https://commons.wikimedia.org/wiki/File:Falling_ball.jpg

Mercator, P. (2013). *Spherical trigonometry* [Digital].
https://commons.wikimedia.org/wiki/File:Spherical_trigonometry_basic_-

triangle.svg

NASA's Scientific Visualization Studio. (2021, January 14). *SVS: Global Temperature Anomalies from 1880 to 2020*. NASA Scientific Visualizations Studio. https://svs.gsfc.nasa.gov/4882

New York Public Library / Science Source / Science Photo Library. (n.d.). *Hipparchus, Greek Astronomer and Mathematician* [Digital]. https://www.sciencephoto.com/media/1011097/view

Temperature Composite. (n.d.). Skeptical Science. Retrieved January 9, 2022, from https://skepticalscience.com//graphics.php?g=7

The Shodor Education Foundation. (2000). *Overview of Computational Science*. ChemViz Curriculum Support Resources. http://www.shodor.org/chemviz/overview/compsci.html

unknown. (1983). *Commemorative Stamp of Muḥammad ibn Mūsā al-K̲wārizmī*. https://commons.wikimedia.org/wiki/File:1983_CPA_5426_(1).png

Urone, P. P., & Hinrichs, R. (2012). 3.4 Projectile Motion. In *College Physics*. OpenStax.

Vogelsberger, M., Genel, S., Springel, V., Torrey, P., Sijacki, D., Xu, D., Snyder, G., Bird, S., Nelson, D., & Hernquist, L. (2014). Properties of galaxies reproduced by a hydrodynamic simulation. *Nature, 509*(7499), 177–182. https://doi.org/10.1038/nature13316

Wicks, R. (2022). *Twelfth night* [Digital]. https://unsplash.com/@robwicks?utm_-source=unsplash&utm_medium=referral&utm_content=creditCopyText

Wing, J. M. (2006). Computational thinking. *Communications of the ACM, 49*(3), 33–35. https://doi.org/10.1145/1118178.1118215

Wing, J. M. (2014, January 10). Computational Thinking Benefits Society. *Social Issues in Computing*. http://socialissues.cs.toronto.edu/index.html%3Fp=279.html

Zhang, Y., & Zhao, Y. (2015). Astronomy in the Big Data Era. *Data Science Journal, 14*(0), 11. https://doi.org/10.5334/dsj-2015-011

2. Computational Thinking and Problem Solving

2.1. Elements of Problem Solving

When faced with a difficult problem that must be solved, we often feel bewildered about how to even get started. While there is no single strategy that can guarantee success, we now understand some systematic approaches to problem solving that can usually get us started.

George Polya, author of *How to Solve It* (Polya, 1971), was a pioneer in the area of heuristics, the formal study of problem-solving. The book *How to Solve It* was addressed to teachers of mathematics with the hope of helping them teach problem-solving to mathematics students, but the model presented in the book is now widely used as a starting point in many other areas of knowledge besides mathematics. The model he presents will be our starting point in developing a general strategy for developing problem solutions that can be used by a computer.

The Polya model consists of four steps:

1. Understand the problem
2. Plan the solution
3. Carry out the plan
4. Look back at the solution

We will reformulate the Polya model to better match development of computer programs. Our working problem-solving model for solving computational problems will have the following general steps:

1. Analysis: Analyze the problem
2. Design: Describe the data and develop algorithms
3. Implementation: Represent data using programming language data struc-

tures and implement the algorithms with specific programming language code

4. Testing: Test and debug the code using test data

2.1.1. Analysis

The analysis stage of problem-solving seeks to gain a fuller understanding of the problem so that possible paths towards a solution can be described. To assist in the analysis of a problem try the following:

■ Restate the problem in your own words.

■ Represent the problem using pictures and diagrams.

■ List the things you know and the things you do not know that are relevant to the problem.

■ Define the goal of your problem solution. Describe what a successful solution would look like.

2.1.2. Design

One of the fundamental methods for designing an algorithm is decomposition: breaking the problem into smaller parts. Decomposition is one of the pillars of computational thinking. The big problem is replaced by a series of smaller problems, each one being more tractable. Sometimes the subproblems are still too complex and decomposition can be applied to them too.

Another element of design is to recognize what kind of data that you have to facilitate a solution and how that data can be described precisely. This process can be characterized as abstraction and data representation.

Ultimately, a major part of design is to develop a specific algorithm for the solution. If you are really stumped by how to get started with your algorithm design, it can be helpful to replace the original problem with a related problem that is easier to solve. After solving the easier problem, you may find that elements of the solution can be incorporated into the original problem solution design. Sometimes you might solve several related problems each of which is easier to solve. Then you can look for patterns in your solutions that might apply to the original problem.

2.1.3. Implementation

The design phase of problem-solving focuses on general descriptions of data representations and algorithms. When the solution requires a computer program, those general design descriptions must be made specific by coding them using a programming language, such as Python. Solution steps described using a flowchart or pseudocode representation will need to be translated into specific Python statements, and data will need to make use of existing data types and structures present in Python or new data structures that can be constructed within the Python language. Performing these steps is the implementation of the problem solution.

2.1.4. Testing

Any problem solution must be evaluated for correctness. When the solution is a computer program such an evaluation can become quite involved. A program that executes without syntax errors is not necessarily providing the output required by the problem solution. This output must be compared with the expectations provided in the analysis phase of our problem-solving strategy. We will discuss some specific ways of testing later once we acquire some additional understanding of coding Python programs.

2.1.5. Example

We will use a physics problem to illustrate using our problem-solving strategy.

Problem Statement: A baseball is hit with an initial velocity of 40.0 [m/s] at an angle of 20°w.r.t. the ground. The ball is 1.2 [m] above the ground when it is hit. How far does the ball travel before hitting the ground?

Analysis: We will start by making some simplifying assumptions.

■ The ball will be modeled as a single particle, so any rotational effects will be ignored.

■ The effects of air resistance will be ignored, so the trajectory can be assumed to be in a single plane.

These assumptions imply the use of abstraction to focus on an initial understanding of the situation. Next, we will draw a picture to fix the geometry, shown in

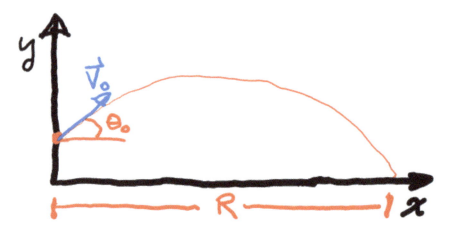

Figure 2.1.: Geometry for the baseball problem.

Figure 2.1.

Using information from the problem statement, we can specify given information.

$$
\begin{aligned}
x_0 &= 0 \\
y_0 &= 1.2\,[\text{m}] \\
v_0 &= 40.0\,[\text{m/s}] \\
\theta_0 &= 20° \\
v_{0x} &= v_0 \cos(\theta_0) = 40.0\,[\text{m/s}] \cos(20°) = 37.5\,[\text{m/s}] \\
v_{0y} &= v_0 \sin(\theta_0) = 40.0\,[\text{m/s}] \sin(20°) = 13.7\,[\text{m/s}]
\end{aligned}
\tag{2.1}
$$

As part of our analysis, we can describe physics laws that might be applicable.

Kinematic Equations for Constant Acceleration:

$$
x = x_0 + v_{0x}t
\tag{2.2}
$$

$$y = y_0 + v_{0y}t - \frac{1}{2}gt^2 \qquad (2.3)$$

The solution to the problem will be the value of R, which is the value of x when $y = 0$.

Design: We can express the plan for calculating R:

1. Put the knowns into Equations 2.2 and 2.3.
2. Solve Equation 2.3 for t when $y = 0$. Call this time T.
3. Substitute the value of T obtained into Equation 2.2 and calculate x.

Implementation: The Design plan requires some algebra, which can be done by using the quadratic formula on Equation 2.3. The result is

$$T = 2.88\,[\text{s}] \Rightarrow R = 108\,[\text{m}] \qquad (2.4)$$

Testing: For this problem solution, testing the solution involves determining whether the value we found is sensible. A distance of 108 [m] is 354 [ft], which is less than the typical distance from home plate to the stadium edge for a professional baseball stadium (~400 [ft]).

Both **Implementation** and **Testing** will look different when the problem involves developing a computer program. We will see many examples of performing these steps as we become more involved in coding.

2.2. Basic Elements of Python

We will be using the Python programming language to develop our skills in computational science. Python was released in 1991 by its primary author, Guido van

Rossum, who named the language after one of his favorite television shows, *Monty Python's Flying Circus*, a comedy series produced by the BBC. Python has gone through many revisions since its introduction. We will use version 3.6 since the Google Colab platform currently supports it.

Here we introduce a few of the basic elements of the Python programming language that will enable us to start implementing our own programs. In particular, we will cover

- Literals
- Variables and Identifiers
- Operators
- Expressions
- Data Types
- Syntax

Extensive documentation for the Python programming language is provided online (Python Software Foundation, 2021).

2.2.1. Literals

One dictionary definition of the word "literal" is "taken at face value" and programming languages use a similar definition: a literal is a sequence of characters that stand for itself. We will consider numeric literals and string literals. These will form building blocks for writing code.

A numeric literal is a character sequence containing only the digits 0-9, a + or − sign, possibly a decimal point, and possibly the letter e, which will indicate scientific notation. A numeric literal with a decimal point is called a **floating-point value**. A numeric literal without a decimal point is called an **integer value**. One additional numerical literal is the **complex value**. It is written as the sum of the real part and imaginary part. The imaginary part is indicated using the character j. Here is an example:

```
c = 3.0 + 4.0j
```

Table 2.1.: Examples of numeric literals

Integer Values	Floating-point Values	Complex Values	Incorrect Values
0	5. , 5.0, +5.0	2.0 + 3.0j	1,500
10	15.25	4 − 5j	1,500.00
25	15.25	0.56 + 1.7j	+1,500
-25	-15.25	-6.0 + 7.0j	-1,500.00

Commas are not used in numeric literals. Table 2.1 gives examples of numeric literals.

A numeric literal with no + or − sign in front of it is considered positive.

Next we consider string literals, usually called "strings". These are sequences of alphanumeric characters. String values are indicated using matching single quotes or double quotes:

'My name is Bob' , "My name is Bob"

Note that a blank character can be in a string value. If the string literal contains an apostrophe, then the value should be enclosed by matching double quotes.

"Bill's Bookstore"

A string literal with no characters in it is called the empty string and is indicated with matching single quotes or double quotes enclosing no characters:

'' or ""

2.2.2. Variables and Identifiers

An **identifier** is a sequence of characters that is used as a name for a programming element such as a function. The rules for constructing identifiers are

They may contain letters and digits but cannot begin with a digit.

39

Table 2.2.: Examples of identifiers.

Valid Identifiers	Invalid Identifiers	Reason
gravitationalforce	gravitationalForce'	Cannot contain quotes
gravitationalForce	2022_year	Cannot start with a digit
gravitational_force	_gravitationalForce	Valid but should be avoided

They are case-sensitive.

They cannot contain quotes

The underscore '_' can be used but should be avoided as the first character.

Table 2.2 gives some examples of identifiers.

A variable is a name assigned to a value. Python assigns a value to a variable by using the assignment operator, =.

```
num = 15
```

The variable num now refers to the value 15. Whenever the variable num is used in a Python expression it will be replaced by the value 15. Consider the expression

```
num + 1
```

This expression will evaluate to 16 since num actually represents the number 15.

A variable can be reassigned to a different value at different points in a program.

2.2.3. Keywords

A **keyword** is a predefined identifier that has a reserved specific use in a programming language. Keywords should not be used as a variable name or other user-defined identifier. If you try to do this then there will be a syntax error when the code is executed. Figure 2.2 shows what happens when you try to use the keyword 'or' as a variable.

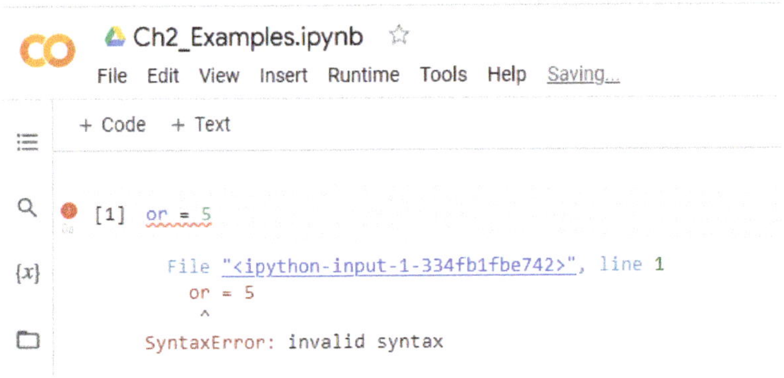

Figure 2.2.: Incorrect use of keyword identifier.

Table 2.3 shows all of the keyword identifiers used in Python 3.6. These identifiers are case-sensitive.

Table 2.3.: Python 3.6 keywords.

and	del	from	None	TRUE
as	elif	global	nonlocal	try
assert	else	if	not	while
break	except	import	or	with
class	FALSE	in	pass	yield
continue	finally	is	raise	
def	for	lambda	return	

2.2.4. Operators

An **operator** is a symbol that indicates some kind of operation performed on one or more operands. An operand is a value, such as a number, that the operator acts on. Arithmetic operators such as addition, subtraction, multiplication, and division require two operands and are called binary operators. An operator that requires only one operand, such as negation, is called a unary operator.

Among the most important operators for scientific and engineering programming are the arithmetic operators shown in Table 2.4.

Table 2.4.: Python arithmetic operators.

Operator	Operation	Example	Result
-x	negation	-100	-100
x + y	addition	100 + 25	125
x - y	subtraction	100 - 25	75
x * y	multiplication	100 * 4	400
x / y	division	100 / 4	25
x // y	truncating division	100 // 3	33
x ** y	exponentiation	10 ** 3	1000
x % y	modulus	50 % 6	2

Notice that unlike in many programming languages, exponentiation uses the operator ** rather than the caret symbol ^.

There will be times when programming requires knowing what data type is produced by an operator. Consider division using the / operator. It will always produce a float, even if both operands are integers, as shown in Figure 2.3.

```
[3]  a = 5 # type int
     b = 2 # type int
     c = a / b
     print(c)
     type(c)

     2.5
     float
```

Figure 2.3.: Data type produced by division.

Now consider truncating division performed by the operator //. If both operands are integers (type int) then the result is an integer. If either operator is a float then the result is a float. Figure 2.4 illustrates this behavior with code snippets.

```
[4]  a = 5 # type int
     b = 2 # type int
     c = a // b
     print(c)
     type(c)

     2
     int
```

(a) Both operands are integer.

```
[8]  a = 5.0 # type float
     b = 2 # type int
     c = a // b
     print(c)
     type(c)

     2.0
     float
```

(b) Top operator is a float.

Figure 2.4.: Data type produced by truncating division.

The modulus operator % may be less familiar than the other arithmetic operators. It is a binary operator that returns the remainder after performing division. If the operands are both integers (type int) then result will be an integer, otherwise the result will be a float.

```
[9]  a = 12 # type int
     b = 5  # type int
     c = a % b
     print(c)
     type(c)

     2
     int
```

(a)

```
[10] a = 12.0 # type float
     b = 5  # type int
     c = a % b
     print(c)
     type(c)

     2.0
     float
```

(b)

Figure 2.5.: Result of modulus operator.

Table 2.5 shows the cyclic nature of the modulus operator.

Table 2.5.: Result of modulo 5 modulus operator.

Operation	Result
0 % 5	0
1 % 5	1
2 % 5	2
3 % 5	3
4 % 5	4
5 % 5	0
6 % 5	1
7 % 5	2
8 % 5	3
9 % 5	4
10 % 5	0

2.2.5. Expressions

An **expression** is a combination of variables, values, and operators that evaluates to a single value. A **sub-expression** is an expression that is part of a larger expression. We will consider arithmetic expressions here. The following are examples of arithmetic expressions in Python.

$$1 + 2, 3 * n + 1, 4 * (n - 1) \tag{2.5}$$

In these expressions, the variable n must have been assigned a value, otherwise, an error will occur.

To avoid ambiguity about the order in which operators are executed in an expression, Python defines a particular order in which operators are applied. Table 2.6 gives the operator precedence used by Python.

Table 2.6.: Python operator precedence. The vertical ordering is from high to low going down.

Operator	Operation	Associativity
**	exponentiation	right-to-left
-	negation	left-to-right
* , / , // , %	mult , division, trunc, div, modulo	left-to-right
+ , -	addition , subtraction	left-to-right

The operator associativity refers to the order in which operators are applied when they have the same precedence. Consider the expression

$$4 ** 3 ** 2 \qquad (2.6)$$

Which exponentiation gets evaluated first: $4 ** 3$ or $3 ** 2$? The right-to-left associativity rule for exponentiation says the $3 ** 2$ operation is performed first.

$$4 ** 3 ** 2 \rightarrow 262144 \text{ not } 4096 \qquad (2.7)$$

If the operator precedence in this situation had been left-to-right then the result of evaluating this expression would have been 4096.

Now consider the expression

$$2/3 * 4 \qquad (2.8)$$

Since the / and * operators have the same precedence we use the left-to-right associativity rule to perform

$$2/3 \rightarrow 0.6666 \text{ then } 0.6666 * 4 \rightarrow 2.6667 \qquad (2.9)$$

yielding the final value 2.6667.

A good practice in coding expressions is to use parentheses to enforce a desired precedence rather than relying on the Python-defined precedence rules. Sub-expressions defined by parentheses are evaluated first.

2.2.6. Data Types

The formal definition of data type is a set of literal values, such as numbers or strings, and the operators that can be applied to them, such as addition and subtraction. Python uses several built-in data types. These are shown in Table 2.7.

Table 2.7.: Python built-in datatypes

Category	Associated Python Keyword
Text Type:	str
Numeric Types:	int, float, complex
Sequence Types:	list, tuple, range
Mapping Type:	dict
Set Types:	set, frozenset
Boolean Type:	bool
Binary Types:	bytes, bytearray, memoryview
None Type:	NoneType

For now, we will focus only on the text and numeric types. Understanding data types can help us avoid problematic operations. For example, division, /, does not apply to strings or mixed string and numeric expressions. The following code will produce an error:

```
s1 = 'This is a string.'
s2 = 'This is another string.'
s1/s2 # This expression will produce an error.
```

Similarly

```
s1 = 'This is a string.'
s1/2 # This expression will produce an error.
```

The built-in function type(v) will tell you the data type of v.

```
[13]  s = 'This is a string.'
      type(s)

      str
```

Figure 2.6.: Example of type( ) function use.

All programming languages use the concept of data type, but the implementation differs between languages. Some languages such as C, C++, or Java require that the data type of a variable be declared explicitly before the variable can be used in an expression. This is called **static typing**. Other languages, including Python, will assign the data type to a variable automatically when the program is executed based on the value referred to by the variable. This is called **dynamic typing**. The same variable name can be assigned a different data type at different places in the program.

A complex expression can have sub-expressions or variables of different data types. These are called **mixed-type expressions**. In order for a mixed-type expression to evaluate to a single value, all the components or operands in the expression must be converted to a single type. This procedure can be done automatically by the Python interpreter, called **coercion**, or the programmer can explicitly perform conversions using data type conversion functions, which is called **type conversion**.

When the Python interpreter performs coercion, it will do so in a way to avoid loss of information. For example in the expression

```
1 + 2.5
```

converting the 1 to 1.0 (int to float) does not result in loss of information but converting 2.5 to 2 (float to int) does. Therefore, the int to float conversion will be done and the final value, 3.5, will be a float. We can perform the type conversions explicitly using type conversion functions. In the above example, explicit type conversion could be done with

$$\text{float}(1) + 2.5 \rightarrow 1.0 + 2.5 \rightarrow 3.5 \qquad (2.10)$$

Table 2.8.: Examples of explicit type conversion.

Type Conversion	Result
float(5)	5
float('5')	5
float('A String')	error
int(5.0)	5
int(5.2)	5
int('5')	5
int('5.2')	error

2.2.7. Syntax

Syntax refers to the structure of a program and the rules about that structure. All languages have syntax. When discussing human languages such as English, syntax refers to the rules of grammar. We will introduce Python syntax rules as we learn about new elements of the language. The Python interpreter will determine if code has syntax errors and will inform you with an error message. Code that is syntactically correct can still generate a runtime error. For example

10/0

is syntactically correct but will generate an error when the expression is evaluated.

A **semantic error** occurs when the program is syntactically correct and generates no runtime errors, but the output is still incorrect. This means that the programmer has not translated the design of the problem solution into code correctly. Semantic errors will be the most difficult ones to eliminate.

2.2.8. Comment Statements

Comment statements are added to code to help humans better understand what is happening in the program. Comment statements are ignored by the Python interpreter, yet they remain an important part of documenting the code. There are two methods of inserting comments:

- Single line comment.
- Comment Block

A single line comment is indicated with the # character. Everything after the # on the line is ignored.

```
# This entire line is ignored
n = 5 # Everything on this line is ignored after the # character
```

A comment block is created with triple single quotes or triple double quotes. All lines between a triple single quote or double quote are ignored until another triple single quote or double quote is encountered.

Single quote block:

```
'''
comment block line 1
comment block line 2
'''
```

Double quote block:

```
"""
comment block line 1
comment block line 2
"""
```

Figure 2.7.: Example of input function use in a notebook.

2.3. Basic Input and Output

When we execute a program interactively, such as in a Jupyter notebook or IPython console, obtaining information from the user and communicating messages to the user will be necessary. The Python built-in functions input and print are simple methods to perform these tasks.

The input function will request information from the user by printing a message and then waiting for the user to enter something using the keyboard. After hitting enter or return the string entered by the user is returned to the program. For example

```
name = input('What is your name?:')
```

will print the string given as the argument of the function, then wait until the user types in something, and then returns what was typed as a string that gets assigned to the variable name. Figure 2.7 shows how this input request looks in a Jupyter notebook.

Basic output is achieved with the print function. It will print items in its argument list to the standard output, which will be an output cell in a Jupyter or Colab notebook. Multiple items can be printed by including them as comma separated items in the function argument list. Figure 2.8 shows an example of this behavior.

The separator between arguments printed out can be specified using the sep keyword argument. The default value is the empty string.

```
[6]  name = input('What is your name?: ')

     What is your name?: Chris
```

```
[7]  print('My name is ', name)

     My name is  Chris
```

Figure 2.8.: Example of printing multiple items.

```
[8]  print('My name is ',name,sep='---')

     My name is ---Chris
```

Figure 2.9.: Use of the sep keyword argument.

If the program needs to obtain a number from the user using the input function then the string that is returned will need to be converted into the appropriate numerical data type using the corresponding type conversion function. The following shows how to obtain a floating point number from the user.

```
float_number = float ( input ( 'Submit a floating point number:') )
```

After execution the variable float_number will contain the number submitted and it will be of type float.

Formatting the printed output of a program can be an important part of the solution. We will discuss here one method for formatting numbers to a specified precision. We will use the built-in format function. Consider the following number.

n = 3.14159

Suppose we want to print n out with 2 decimal places. We can do this with the format function, which has the following syntax:

```
format(value, format_specifier)
```

where

format_specifier: '[width][.precision]type'

width – total number of characters in the field

precision – number of decimal places

type –
  f – float
  e – exponential

Note that the width and precision fields are optional, which is indicated with the brackets.

To format the variable n to two decimal places, use the following code.

```
[13] n = 3.14159
     s = format(n,'5.2f')
     print(s)

     3.14
```

Figure 2.10.: Example of using the format function.

2.4. Exercises

1. The Polya general problem solving model is reformulated for computational problems to consist of the following four steps (select the correct ones):

  a. Analysis
  b. Design

c. Implementation

d. Testing

e. Debugging

f. Flowcharting

2. Which of the following are correctly formulated numerical literals?

a. 1,200

b. 1200

c. 1200.

d. 1200.0

3. Which of the following are legal Python variable names?

a. 3CPO

b. n1

c. 'number'

d. chiefExecutive

e. for

4. What value will the following expression have?

3 + 8 * (3**2 - 8) / 10

a. 3.0

b. 3

c. 3.8

d. 4

5. What is the value produced by each of the following expressions?

a. 12/5

b. 12./5.

c. 12.0//5.0

d. 12%5

6. What is the data type of the variable n after the following code has been executed?

```
n = 2
n = n*3
n = n/3
```

7. What is the data type of the variable num after the following statement is executed?

```
num = input('Type in a floating point number: ')
```

a. int

b. float

c. str

2.5. Program Modification Problems

1. The code presented below will request that the user submit a temperature in Fahrenheit and then converts it to a temperature in Celsius and prints out the result.

```
"""
Program: Temperature Conversion Program (Fahrenheit to Celsius)
Author: C.D. Wentworth
Version: 1.25.2020.1
Summary: This program will convert a temperature entered in Fahrenheit
         to the equivalent temperature in Celsius
```

```
"""

# get temperature in Fahrenheit
TF = float(input('Enter degrees Fahrenheit: '))

# calc degrees Celsius
TC = (TF - 32) * 5 / 9

# output degrees Celsius
print(TF, 'degrees Fahrenheit equals',
      format(TC, '.1f'), 'degrees Celsius')
```

Revise the program to do the following:

- Print out a program greeting that describes what the program will do.
- Requests that the user submit a temperature in Celsius.
- Prints out the temperature in Fahrenheit with an appropriate message.

2. Create a program that requests a temperature from the user in Fahrenheit and then prints out the temperature in Celsius and Kelvins with appropriate messages. You can start with the code from Program Modification Problem 1.

2.6. References

Polya, G. (1971). *How to Solve It: A New Aspect of Mathematical Method* (2nd ed.). Princeton University Press.

Python Software Foundation. (2021). *Python 3.6.15 Documentation.* https://docs.python.org/3.6/

3. Control Structures

Motivating Problem: Temperature Conversions

Scientists and engineers must use several temperature scales in their work. In the United States, most of the public still uses the Fahrenheit scale, while in scientific research the metric scales Celsius and Kelvin are preferred. Modern thermodynamic science postulates a lowest possible temperature for a physical system, which suggests using temperature scales for which 0 corresponds to this lowest possible temperature. The Kelvin scale is the absolute temperature scale in the metric system. Engineers working in the U.S. must often use the Rankine temperature scale, which associates 0 with absolution zero, but where a temperature difference of one Rankine equals one degree Fahrenheit.

Given the need for scientists and engineers to use different temperature scales, creating a computer program that can take a temperature given in one scale by a user and convert it to a temperature in another scale is an important problem. We will need additional ways to control the flow of a program to solve this problem, in particular, we must execute different blocks of code depending on which temperature conversion must be done.

This chapter introduces elements of the programming language that allow for sophisticated program control based on information submitted by the user. We will learn about Boolean expressions and how they are used by the different control structures available in Python.

3.1. Definition of Control Structures

Control flow refers to the sequencing of executed statements in a program. A control structure is a type of statement in a programming language that determines the control flow in a program. All programming languages provide the following control structures:

- sequential control
- selection control
- iterative control

Sequential control is the execution of statements one after the other. It is what we have used previously. **Selection control** involves executing a different block of code depending on the value of a conditional statement, such as one that evaluates to true or false. **Iterative control** allows for the repetition of a block of statements based on a condition. Figure 3.1 illustrates these three kinds of control structures.

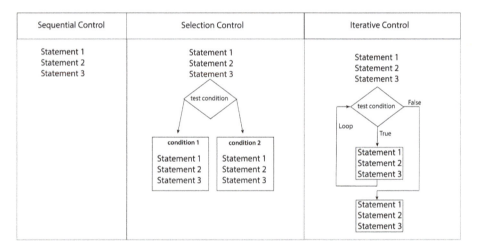

Figure 3.1.: The three kinds of flow control structures.

We see in Figure 3.1 that sequential control involves executing a statement, then executing the next statement, and so on. Selection control shows the program arriving at a decision point where some kind of condition statement is evaluated, often a true/false kind of statement, then the program flow goes to a different code block depending on the value of the condition. Iterative control shows the program arriving at a decision point where a condition statement is evaluated and as long as the condition is true then a code block gets executed in a loop, and when the condition becomes false the program control continues after the loop.

3.2. Boolean Expressions

Selection and iterative control often involve a logical expression that evaluates to true or false. Such values are called Boolean data type. Programming languages use Boolean algebra to evaluate expressions involving Boolean variables or values. This type of algebra was developed by the 19th century mathematician George Boole . Instead of operations performed on numbers, as is done in traditional algebra, Boolean algebra involves operations performed on True/False values. We will introduce the Boolean (or logical) operators **and**, **or**, and **not**, and learn how to use them to construct complex Boolean expressions, which always evaluate to either True or False.

In Python, the logical values of true and false are indicated by True and False. The capitalization is important. Also note that the Boolean operators are keywords in Python, so they should not be used as variable names and other identifiers.

The **and** and **or** operators are binary operators requiring two Boolean variables or values as operands. The **not** operator is a unary operator, which simply turns the logical value to its opposite: True becomes False and False becomes True. The result of Boolean operators is defined by the Boolean Truth Table given in Table 3.1.

Table 3.1.: Boolean logic truth table.

x	y	x and y	x or y	not x
FALSE	FALSE	FALSE	FALSE	TRUE
TRUE	FALSE	FALSE	TRUE	FALSE
FALSE	TRUE	FALSE	TRUE	TRUE
TRUE	TRUE	TRUE	TRUE	FALSE

Relational operators are often involved in Boolean expressions. These operators These operators are used in comparing two values. The Python relational operators are shown in Table 3.2.

Table 3.2.: Python relational operators.

Operator	Meaning	Example	Result
==	equal	15 == 20	FALSE
!=	not equal	15 != 20	TRUE
<	less then	15 <20	TRUE
>	greater than	15 >20	FALSE
<=	less than or equal	15 <= 15	TRUE
>=	greater than or equal	15 >= 20	FALSE

The relational operators can be applied to any kind of value that has an ordering, including strings. The ordering of string values follows a dictionary ordering based on a particular way of assigning numbers to a character. Python uses the Unicode Standard (Unicode Consortium, 2021) for encoding characters. In this system 'A' is less than 'B', because the Unicode code for 'A' is 65 and the Unicode code for 'B' is 66. Since the Unicode code for lowercase letters is greater than it is for uppercase letters we have 'a' is greater than 'B'.

We should be careful to avoid errors in using relational operators coming from not distinguishing between meaningful mathematical expressions and meaningful programming language expressions. Consider the following.

$$10 <= num <= 20 \tag{3.1}$$

Assume that the variable num represents the number 15. The expression is understood in mathematics to mean

$$(10 <= num) \text{ and } (num <= 20) \tag{3.2}$$

which evaluates to True, but most programming languages will not make this implicit assumption. Instead the evaluation will proceed as follows.

$$10 <= \text{num} <= 20 \rightarrow (10 <= \text{num}) <= 20 \rightarrow \text{True} <= 20 \rightarrow \text{error} \quad (3.3)$$

The final step yields an error since we are trying to compare two different kinds of data, Boolean and numerical. To avoid this problem, you should use parentheses explicitly, as shown in Equation 3.2.

The Boolean operators must be added to the operator precedence definitions. Table 3.3 gives the operator precedence order including Boolean operators.

Table 3.3.: Updated operator precedence table.

Operator	Operation	Associativity
**	exponentiation	right-to-left
-	negation	left-to-right
* , / , // , %	mult , division, trunc, div, modulo	left-to-right
+ , -	addition , subtraction	left-to-right
<,>,<=,>=,!=,==	relational operators	left-to-right
not	not	left-to-right
and	and	left-to-right
or	or	left-to-right

Parentheses can be used in expressions to help humans reading the code to recognize the desired operator precedence. Consider the expression

$$1 + 2 < 3 + 4$$

The expression is evaluated using the operator precedence order in Table 3.3 as

$$1 + 2 < 3 + 4 \rightarrow 3 < 7 \rightarrow \text{True}$$

To assist a human reader of the code the operator precedence could be made explicit by writing the expression as

$$(1 + 2) < (3 + 4)$$

Using parentheses in this way is good programming practice.

3.3. Selection Control Overview

Selection control allows for different blocks of statements to be executed depending on the value of a conditional expression, often a Boolean expression. We will discuss the Python if statement as the primary example of this control structure. Here is the basic form of the if statement:

```
if condition:
    statement block
else:
    statement block
```

The condition in this statement must be a Boolean expression, that is one that evaluates to True or False. Also, the else part is optional. Here is a specific example.

```
if grade >= 70:
    print('satisfactory performance')
else:
    print('improvement needed')
```

One thing to note about Python syntax is that indentation is important. All statements in a particular statement block must be indented the same amount. To clarify this situation, we must introduce some additional vocabulary. Consider a simplified version of the temperature conversion task. Suppose that the problem

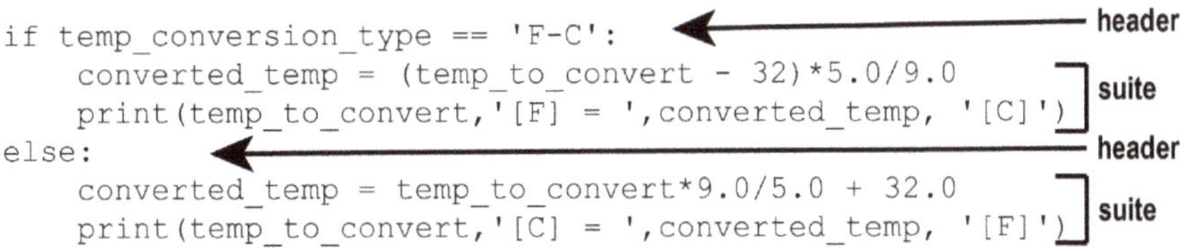

```
if temp_conversion_type == 'F-C':                          header
    converted_temp = (temp_to_convert - 32)*5.0/9.0      ⎤
    print(temp_to_convert,'[F] = ',converted_temp, '[C]')⎦ suite
else:                                                       header
    converted_temp = temp_to_convert*9.0/5.0 + 32.0      ⎤
    print(temp_to_convert,'[C] = ',converted_temp, '[F]')⎦ suite
```

Figure 3.2.: Control structure vocabulary.

is to convert Fahrenheit to Celsius or Celsius to Fahrenheit, depending on what the user specifies. Assume that the user inputs F-C or C-F depending on which conversion must be done. This code will be contained in temp_conversion_type. The temperature to be converted will be in the variable temp_to_convert. The converted temperature will be in the variable converted_temp. Figure 3.2 introduces some vocabulary to help in discussing control structures.

The **header** in a control structure is a keyword and ends in a colon. The if statement has two headers. Headers in the same control structure must be indented the same, as seen in 3.2. A suite or block is a set of statements following a header. They must all be indented the same. The following code will generate an error.

```
if temp_conversion_type == 'F-C':
        converted_temp = (temp_to_convert - 32)*5.0/9.0
    print(temp_to_convert,'[F] = ', converted_temp, '[C]')
else:
    converted_temp = temp_to_convert*9.0/5.0 + 32.0
    print(temp_to_convert,'[C] = ' , converted_temp, '[F]')
```

The amount of indentation of a code block does not matter, although four spaces has become a convention for Python.

Multi-way selection can be achieved by using nested if statements or by using the elif header. Nested if statements have the following structure.

```
if condition:
    statement block
else:
    if condition:
        statement block
    else:
        if condition:
            statement block
        else:
            statement block
```

Figure 3.3.: Multi-way selection with nested if statements.

The same flow control is achieved more cleanly using the elif header shown in Figure 3.4.

```
if condition:
    statement block
elif condition:
    statement block
elif condition:
    statement block
else:
    statement block
```

Figure 3.4.: Multi-way selection using the elif header.

3.4. Iterative Control Overview

Iterative control structures allow a block of code to be repeated based on a condition that appears in the header for the structure. As Figure 3.1 suggests, an iterative control structure appears as a loop due to the repetitive nature of the flow. The first iterative control structure that we will introduce is the while loop, which repeatedly executes a code block based on a Boolean expression. The syntax for a while loop is

```
while condition:
    statement block
```

The condition must be a Boolean expression that evaluates to True or False. As long as the condition is true, the block of code will execute. When the condition becomes False, the flow control jumps to the first statement after the while loop. Here is a simple example of using the while loop to achieve iterative control:

```python
num = 0
count = 1
number_of_iterations = int(input('Enter the number of iterations:'))
while count <= number_of_iterations:
    num = num + 1
```

All iterative flow control can be achieved with the while loop, but if the number of iterations is known in advance, then another control structure can be used: the for loop. When this structure is used the loop is executed for each element of a given sequence. We will discuss Python sequence data types in the next chapter, but a commonly used example is provided by the range function.

The range function will produce a sequence of integers that can be used by a for loop. The syntax is

```python
range(start, stop, step)
```

where

start - An integer number specifying at which position to start the integer list. Default is 0. Optional

stop - An integer number specifying at which number to stop. The sequence goes up to but not including this number. Required.

step - An integer number specifying the increment. Default is 1. Optional.

Examples:

```python
range(2,5)
```

generates the sequence

2,3,4

```
range(5)
```

generates

0,1,2,3,4

```
range(0,10,2)
```

generates

0,2,4,6,8

The syntax for a for loop is shown below.

```
for i in sequence:
    statement block
```

The following example uses the range function to generate a sequence of integers that can be used in a for loop.

```
for i in range(1,6):
    print('Index value is',i)
```

This for loop will produce the following output

```
Index value is 1
Index value is 2
Index value is 3
Index value is 4
Index value is 5
```

3.5. Pseudocode

The Design phase of solving a computational problem can be challenging since it is where we must develop the actual algorithm for the solution. By recognizing the key logical structures of programming without concern with syntax details of a specific programming language we can often work through the basic logic of the solution. One method used by software engineers to construct an expression of this logic is writing pseudocode.

Pseudocode involves using plain language to write out the basic logic of the problem solution using structural conventions of programming languages but not being concerned with specific syntax details. Pseudocode provides a bridge between initial thoughts about a problem solution and an actual working code.

Structural elements of programming languages that we want to recognize in our pseudocode version of the program include control structures and functions. We will use a particular style of writing the pseudocode that recognizes these structural elements.

Let's start with control structures. There three major kinds of program control: sequential, selection, and iterative. Sequential control involves executing a step and then moving on to the next step, one after the other. In our pseudocode we will always begin a step with a capital letter, as in the following:

```
Step 1
Step 2
Step 3
.
.
.
```

Figure 3.5.: Sequential control in pseudocode.

Selection control involves executing different blocks of code depending on the value of a Boolean conditional variable. This will involve using an if/else type of structure. Our pseudocode style for indicating a selection control block is to make the keywords IF and ELSE uppercase. We will also indicate the end of the IF/ELSE block with the keyword ENDIF.

```
IF condition
    Step 1
    Step 2
ELSE
    Step 1
    Step 2
ENDIF
```

Figure 3.6.: Selection control in pseudocode.

Iterative control involves setting up a loop, which can be done with the while structure or a for loop. Our pseudocode style will require making the appropriate keyword uppercase. The block will end with the keyword ENDWHILE. Here is how the while structure will be indicated:

```
WHILE condition DO
    Step 1
    Step 2
ENDWHILE
```

Figure 3.7.: Iterative control in pseudocode. While loop.

The for loop will be written as

```
FOR item IN list DO
    Step 1
    Step 2
ENDFOR
```

Figure 3.8.: Iterative control in pseudocode. For loop.

When we use decomposition and abstraction to help solve a computational problem, we often write separate program units to take care of a particular step in the algorithm. In Python, these program units are typically functions. When writing pseudocode and we decide to make a step of the algorithm into a separate function then we will write a pseudocode version of the function as a separate program unit using the FUNCTION keyword. Important parts of writing the pseudocode for a function are to identify the required parameters passed to the function and the data that will be returned to the main program. We will use the following for the pseudocode style.

```
FUNCTION function_name
    INPUT:parameter1, parameter2
    Step 1
    Step 2
    .
    .
    .
    OUTPUT return_variable1, return_variable2
ENDFUNCTION
```

Figure 3.9.: Function definition in pseudocode.

We will indicate the main program in our pseudocode using the PROGRAM keyword.

```
PROGRAM program_name
    Step 1
    Step 2
    .
    .
    .
ENDPROGRAM
```

Figure 3.10.: Program unit in pseudocode.

As an example of developing a pseudocode version of an algorithm, let us consider the problem of cleaning up the yard of a residential property. We start with a high-level view of the algorithm.

```
PROGRAM clean_up_the_yard
    Inspect the yard for debris, excessive grass height, \
        unkempt hedges
    IF there is debris
        Pick up debris
    ENDIF
    IF there is excessive grass height
        Mow the lawn
    ENDIF
    IF there are unkempt hedges
        Trim the hedges
    ENDIF
ENDPROGRAM
```

Figure 3.11.: Highlevel pseudocode version of the clean_up_the_yard algorithm.

Now we will produce a more detailed version of the algorithm, as shown in Figure 3.12. Note that for each of the steps listed in the pseudocode of Figure 3.11 we add additional detailed steps.

3.6. Computational Problem Solution: Temperature Conversion

We return to the temperature conversion problem described at the beginning of this chapter. We will develop a program that solves this problem using programming elements developed in the chapter.

3.6.1. Analysis

The first step in problem analysis is to state the problem precisely and describe how to recognize a successful solution.

Problem Statement: The program requests that the user submit a temperature including scale used and also requests to which scale the temperature should be converted. The Program should perform some error detection to ensure that the submitted temperature is not below absolute zero on the user chosen scale. The program will print out the converted temperature with appropriate label. Temper-

```
PROGRAM clean_up_the_yard
    Inspect the yard for debris, excessive grass height, \
        unkempt hedges
    IF there is debris
        Determine the kind of debris
        IF debris is mainly leaves
            Get some 30 gallon yard bags
            Rake up leaves
            Put leaves in bags
        ELSE
            Get the wheelbarrow
            WHILE debris remains DO
                Fill up wheelbarrow
                Empty wheelbarrow at wood pile
            ENDWHILE
        ENDIF
    ENDIF
    IF there is excessive grass height
        Get the lawn mower
        Check the gas
        IF mower needs gas
            Fill the gas tank
        ENDIF
        WHILE the lawn is not finished DO
            Mow a lap
        ENDWHILE
    ENDIF
    IF there are unkempt hedges
        Get the hedge clippers
        Trim the hedges
    ENDIF
ENDPROGRAM
```

Figure 3.12.: A more detailed version of the clean_up_the_yard algorithm.

ature scales that will be used are Celsius, Fahrenheit, and Kelvin. Temperatures should be formatted to show two digits right of the decimal point.

Part of our analysis is to research physical principles that may be helpful in solving the problem. In this case, we can look up absolute zero for each of the scales being considered. These temperatures are shown in Table 3.4.

We continue our background research to find the equation for each of the required conversions. Equations 3.4-3.9 show the required conversion equations.

Table 3.4.: Absolute zero for the temperature scales considered.

Scale	Absolute Zero
Celsius	-273.15
Fahrenheit	-459.67
Kelvin	0

$$T_C = T_K - 273.15 \tag{3.4}$$

$$T_C = (T_F - 32) \times \frac{5}{9} \tag{3.5}$$

$$T_F = \frac{9}{5} \times T_C + 32 \tag{3.6}$$

$$T_F = (T_K - 273.15) \times \frac{9}{5} + 32 \tag{3.7}$$

$$T_K = T_C + 273.15 \tag{3.8}$$

$$T_K = (T_F - 32) \times \frac{5}{9} + 273.15 \tag{3.9}$$

These equations can be used to generate some test data that will be used to with our program. This allows us to recognize a program that works correctly. Table 3.5 shows relevant test data.

3.6.2. Design

The solution design for our problem should include a description of the algorithm using pseudocode and a list of the required data structures (variable names and types) used in the code. Table 3.6 shows the required variables for implementing the algorithm for this problem solution. Typically, we would start this table as we begin to think about the algorithm and then add to it as the algorithm develops.

Python uses a preferred style for creating variable names (Rossum et al., 2022). This style recommends that variable names (and function names) be all lower case with individual words in the name separated by an underscore. We should note that other programming languages use different style conventions and companies that use Python for software development will sometimes have a different style guide. We will use the PEP 8 recommended style in this book.

Table 3.5.: Test data for the temperature conversion program.

Submitted Temperature		Converted Temperature	
T	Scale	T	Scale
-300	C	Error	
-150	C	-238	F
-150	C	123.15	K
0	C	32	F
0	C	273.15	K
-500	F	Error	
-400	F	-240	C
-400	F	33.15	K
0	F	-17.78	C
0	F	255.37	K
-100	K	Error	
0	K	-273.15	C
0	K	-459.67	F
150	K	-123.15	C
150	K	-189.67	F

After starting the list of required data structures (variable names, in this case), we write out a high-level pseudocode version of the algorithm, as shown in Figure 3.13.

Table 3.6.: Required data structures for the algorithm.

Data Structure	Type	Description
submission_is_incorrect	Boolean variable	specifies whether submitted temperature is above absolute zero
submitted_temperature	float variable	temperature submitted by user
submitted_scale	string variable	scale for submitted_temperature
converted_temperature	float variable	the converted temperature
converted_scale	string variable	scale for converted_temperature

```
PROGRAM temperature_conversion
    Print a program greeting
    submission_is_incorrect = True
    WHILE submission_is_incorrect DO
        Request that the user submit a temperature to convert (defines
          submitted_temperature)
        Request that the user submit the scale of submitted
          temperature (defines submitted_scale)
        Check whether submitted_temperature is above absolute zero
        IF submitted_temperature is above absolute zero
            submission_is_incorrect = False
        ENDIF
    ENDWHILE
    submission_is_incorrect = True
    WHILE submission_is_incorrect DO
        Request that the user submit a scale for the conversion
          (defines convertedScale)
        IF convertedScale is okay
            submission_is_incorrect = False
        ENDIF
    ENDWHILE
    Calculate converted_temperature
    Print out converted_temperature
ENDPROGRAM
```

Figure 3.13.: Pseudocode version of the problem solution algorithm.

Note that the while loops implement our error detection for submitted input.

3.6.3. Implementation

We will now create a Python code implementation of each part. The program greeting and first while loop that checks for a correct temperature submission is shown in Figure 3.14. Note that the program code starts with a brief documentation comment, which includes a few key items such as the author, a version number, and a brief summary of what the program does. This is a good step towards providing documentation for the program. In addition to the iterative control structure (while loop), this section uses the mult-way selection structure (if elif).

```python
"""
Program Name: Temperature Scale Conversion
Author: C.D. Wentworth
version: 6.23.2022.1
Summary: Temperature Conversion Program with Input error checking of
         temperatures. This version accepts a user-defined temperature in
         Celsius, Fahrenheit, or Kelvin and converts the temperature
         to a user selected scale.
"""
# Display program greeting
print('Welcome to the Temperature Scale Conversion Program!')
print('This program will request that the user submit a temperture.')
print('Next, it requests the converted scale.')
print('Finally, it prints out the converted temperature.')
submission_is_incorrect = True
scale_request = "'C' for Celsius, 'F' for Fahrenheit, 'K' for Kelvin "
while submission_is_incorrect:
    submitted_temperature = float(input('Submit a temperature to convert: '
))
    print('Specify the scale of your submitted temperature: ')
    submitted_scale = input(scale_request)
    if ((submitted_scale == 'C') and (submitted_temperature>=-273.15)):
        submission_is_incorrect = False
    elif ((submitted_scale == 'F') and (submitted_temperature>=-459.67)):
        submission_is_incorrect = False
    elif ((submitted_scale == 'K') and (submitted_temperature>=0)):
        submission_is_incorrect = False
    else:
        print('Incorrect submitted temperature. Try again.')
```

Figure 3.14.: Program greeting and temperature request.

Next, we code the program part that requests that the user submit a temperature scale for the converted temperature. The while loop provides for some user input error detection.

```
29 print('What scale should be used for the converted temperature?')
30 converted_scale = input(scale_request)
31 submission_is_incorrect = True
32 while submission_is_incorrect:
33     if ((submitted_scale == 'C') and
34         (converted_scale == 'F' or converted_scale == 'K')):
35         submission_is_incorrect = False
36     elif ((submitted_scale == 'F') and
37         (converted_scale == 'C' or converted_scale == 'K')):
38         submission_is_incorrect = False
39     elif ((submitted_scale == 'K') and
40         (converted_scale == 'C' or converted_scale == 'F')):
41         submission_is_incorrect = False
42     else:
43         print('There is a problem with your submission.')
44         converted_scale = input(scale_request)
```

Figure 3.15.: Converted temperature scale request.

Finally, we code the actual temperature conversion calculation and print out the result.

```python
45  # Calculate the converted temperature.
46  if submitted_scale == 'C':
47      if converted_scale == 'F':
48          # convert Celsius to Fahrenheit
49          converted_temperature = submitted_temperature*9.0/5.0 + 32.0
50      else:
51          # convert Celsius to Kelvin
52          converted_temperature = submitted_temperature + 273.15
53  elif submitted_scale == 'F':
54      if converted_scale == 'C':
55          # convert Fahrenheit to Celsius
56          converted_temperature = (submitted_temperature - 32.0)*5./9.
57      else:
58          # convert Fahrenheit to Kelvin
59          converted_temperature = (submitted_temperature - 32.0)*5.0/9.0 +
        273.15
60  else:
61      if converted_scale == 'C':
62          # convert Kelvin to Celsius
63          converted_temperature = submitted_temperature - 273.15
64      else:
65          # convert Kelvin to Fahrenheit
66          converted_temperature = (submitted_temperature - 273.15)*9.0/5.0 +
        32.0
67  # print out the result
68  s1 = format(submitted_temperature,'.2f')
69  s2 = format(converted_temperature,'.2f')
70  print(s1,submitted_scale,' is ',s2,converted_scale)
```

Figure 3.16.: Temperature calculation part.

3.6.4. Testing

Table 3.5 gives some test data to use with the program. We reproduce that table below (Table 3.7) and add a column giving the actual program output. Since the output column matches the expected converted temperature column we have some evidence that the program runs correctly.

Table 3.7.: Test data with program output.

Submitted Temperature		Converted Temperature		
T	Scale	T	Scale	Program Output
-300	C	Error		error detected
-150	C	-238	F	-238
-150	C	123.15	K	123.15
0	C	32	F	32
0	C	273.15	K	273.15
-500	F	Error		error detected
-400	F	-240	C	-240
-400	F	33.15	K	33.15
0	F	-17.78	C	-17.78
0	F	255.37	K	255.37
-100	K	Error		error detected
0	K	-273.15	C	-273.15
0	K	-459.67	F	-459.67
150	K	-123.15	C	-123.15
150	K	-189.67	F	-189.67

3.7. Exercises

1. Describe the three kinds of program flow control.

2. What is the value of each of the following Boolean expressions?

   a. 10 >= 20

   b. 10 != 20

   c. (10 >= 20) and (10 != 20)

   d. (10 >= 20) or (10 != 20)

   e. 'Jim'<'Mike'

   f. 12*2 == 8*3

3. Correct the indentation in the following code block so that it will execute without an error.

```
score = 81
if score >= 90:
    print('Your score indicates outstanding work.')
elif score >=80:
    print('Your score indicates good work.')
      print('You can resubmit your work.')
elif score >=70:
    print('Your work is adequate.')
    print('Try to improve your score.')
else:
    print('Your work needs improvement.')
    print('Please try again.')
```

4. Develop an algorithm, and pseudocode implementation of it, for the problem of cleaning an apartment with multiple bedrooms. Produce two versions: a high-level version with not many details, and then a more detailed version.

3.8. Program Modification Problems

1. In this exercise you are given a program (shown below) that obtains two integers from the user and then prints a message depending on whether the integers are both zero or not. You must change the code so that it tests whether both integers are zero, one is zero and one is non-zero, or both are non-zero and prints an appropriate message.

```
"""
Program Name: Chapter 3 Prog Mod Prob 1
Author: C.D. Wentworth
version: 6.23.2022.1
Summary: This program requests that the user
         submit two integers and then
         determines whether one is zero or
         not and then prints a message.
"""
# Ask user for two integers
i1 = input('type in an integer ')
i1 = int(i1)
i2 = input('type in another integer ')
i2 = int(i2)
if i1 == 0 or i2 == 0:
    print('One of the integers is zero')
```

```
else:
    print('Neither integer is zero')
```

2. Add the Rankine temperature scale to the available scales in the Temperature Scale Conversion Program. Remember to check that the temperature submitted by the user is greater than or equal to absolute zero.

3.9. References

Rossum, G. van, Warsaw, B., & Coghlan, N. (2022). *PEP 8 – Style Guide for Python Code | peps.python.org.* https://peps.python.org/pep-0008/

Unicode Consortium. (2021). *Unicode.* Unicode. https://home.unicode.org/

4. Lists

Motivating Problem: Identifying Patterns in Global Surface Temperature Data

Climate science generates large amounts of data, including temperature and solar irradiance at specific locations on the earth's surface over many years and time series of greenhouse gas concentrations in the atmosphere. Identifying patterns in the data is one of the important challenges for climate scientists. In this chapter, we will develop some of the computational science skills that will help with this challenge.

Consider one crucial climate observation: the average surface temperature of the earth. Table 4.1 shows some of the surface temperature data produced by the NASA Goddard Institute for Space Studies Surface Temperature Analysis group (GISTEMP Team, 2022b). The complete data set covers the years 1880-2021.

If you look at the entire 1800-2021 data table and try to identify patterns, you will likely be challenged, but a simple scatter graph of the annual mean quickly shows an interesting pattern, as seen in Figure 4.1.

Table 4.1.: GLOBAL Land-Ocean Temperature Index. The numbers are the surface temperature anomaly multiplied by 100 degrees Celsius using 1951-1980 as the base period. The full dataset goes through 2022.

Year	Jan	Feb	Mar	Apr	May	Jun	Jul	Aug	Sep	Oct	Nov	Dec	Ann J-D	Ann D-N	DJF	MAM	JJA	SON	Year
1880	-18	-24	-9	-16	-10	-21	-18	-10	-14	-24	-22	-18	-17	***	****	-12	-16	-20	1880
1881	-20	-14	3	5	6	-19	0	-3	-15	-22	-18	-7	-9	-10	-17	5	-7	-18	1881
1882	16	14	4	-16	-14	-22	-16	-8	-15	-23	-17	-36	-11	-9	8	-8	-15	-18	1882
1883	-29	-37	-12	-19	-18	-7	-7	-14	-22	-11	-24	-11	-18	-20	-34	-16	-9	-19	1883
1884	-13	-8	-36	-40	-33	-35	-33	-28	-27	-25	-33	-31	-28	-27	-11	-37	-32	-28	1884
1885	-58	-33	-26	-42	-45	-43	-33	-31	-29	-23	-24	-10	-33	-35	-41	-38	-36	-25	1885
1886	-44	-51	-43	-28	-24	-35	-18	-31	-24	-28	-27	-25	-31	-30	-35	-32	-28	-26	1886
1887	-72	-57	-35	-35	-30	-25	-26	-35	-26	-35	-26	-33	-36	-36	-51	-33	-29	-29	1887
1888	-34	-36	-41	-20	-22	-17	-10	-14	-12	2	3	-4	-17	-20	-34	-28	-14	-2	1888
1889	-9	17	6	10	-1	-10	-8	-20	-24	-25	-33	-29	-10	-8	1	5	-13	-27	1889
1890	-42	-44	-40	-30	-39	-24	-28	-39	-36	-25	-43	-31	-35	-35	-38	-36	-30	-35	1890

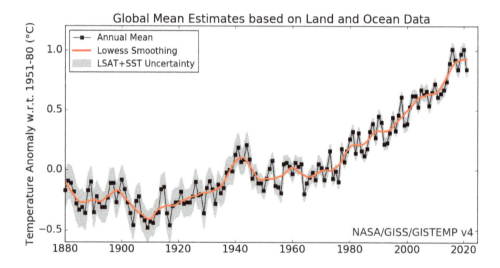

Figure 4.1.: Global mean temperature anomaly (GISTEMP Team, 2022a).

With the help of concepts developed in this chapter, the problem we must solve is how to create a graph, as shown in Figure 4.1, using tables of numerical data such as that in Table 4.1. Solving this problem will require understanding the sequence data type, reading data from a text file, and the basic use of a plotting library. We will learn about the specific examples of sequences in Python, including lists and numpy arrays. We will briefly introduce the Matplotlib plotting library used extensively for Python-based scientific visualizations.

4.1. General Properties of Lists

Lists are found frequently in everyday life: grocery lists, to-do lists, and life goals for the year. Science and engineering work also requires the use of lists. The yearly global surface temperature data introduced in Table 4.1 can be represented as a list. Work in genetics often requires knowing the sequence of nucleotides in a molecule of DNA, and such a sequence can be thought of as a list.

Work with lists computationally will be expedited by having an appropriate data structure representing the list structure. Python uses several such data structures. We will discuss these structures by first introducing the sequence idea, which can be considered a general definition of a list. A **sequence** is a linearly ordered set of values, each of which can be accessed using an index number. Table 4.2 gives an example of a sequence. The first value, -33, is indexed by 0 since Python uses a **zero-based indexing** method. Some programming languages would start indexing with the numeral 1, but not Python. This is a crucial point to keep in mind as you work with sequence data structures in Python.

Table 4.2.: Example of a sequence.

index	Value
0	-33
1	-12
2	4
3	-37
4	-19

Several operations can be performed on most sequence data structures. These operations include retrieving, updating, inserting, removing, and appending an element in the sequence. Figure 4.2 illustrates the retrieve operation. The original sequence remains unchanged.

Updating a sequence value changes the value stored at a particular location. The

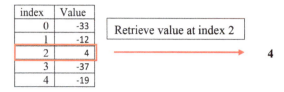

Figure 4.2.: Retrieve sequence operation.

index	Value
0	-33
1	-12
2	4
3	-37
4	-19

Update value at index 2 with 6

index	Value
0	-33
1	-12
2	6
3	-37
4	-19

Figure 4.3.: Update sequence operation.

original value is lost, and the sequence keeps the same length. Figure 4.3 illustrates this operation.

The **insert** operation places a new value at a particular position specified by the index but shifts all other values down. No values are lost, but the sequence increases in length. Figure 4.4 illustrates this effect.

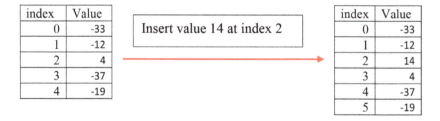

Figure 4.4.: Insert sequence operation.

The **remove** operation takes out the value at a specific index and then shifts all the following up in the sequence, as shown in Figure 4.5.

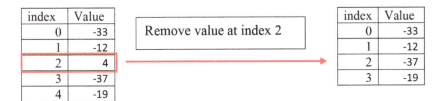

Figure 4.5.: Remove sequence operation.

The append operation attaches a new value to the end of the sequence, as shown in Figure 4.6.

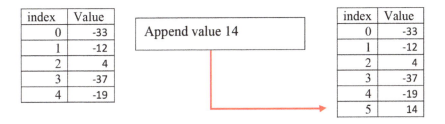

Figure 4.6.: Append sequence operation.

Sequences can be categorized as mutable, meaning that they can be changed, or immutable, meaning that the sequence cannot be changed after being defined. An immutable sequence can only have the retrieve operation applied to it. The built-in Python sequence data types include Python lists, tuples, and strings.

4.2. Python Lists

Python features a built-in sequence data structure called a list. A Python list is a mutable sequence, and the elements can have mixed data types. A Python list does not need to be all floats or all strings. It can contain a combination of floats, strings, and any other valid Python data type, including a Python list. A Python list is denoted using square brackets with elements separated by commas. Here are some examples of Python lists:

```
[50.5, 12.4, 72.5]      [2, 4, 'two', 'four']
```

An empty list is indicated using an empty pair of square brackets:

```
[]
```

A Python variable name can be assigned to a list to facilitate working with it.

```
list_example = [10, 20, 30, 40]
```

The syntax for accessing individual elements of a list (and other sequence types) is to use the index operator [] (not the same as the empty list) with an expression representing an index number inside the brackets.

```
list_example[0]  ───────▶  10   first element

list_example[1]  ───────▶  20   second element

list_example[2]  ───────▶  30   third element
```

Remember that Python uses zero-based indexing, so the first element of a sequence is referenced with 0. Negative index values reference elements from the right or bottom of the list rather than from the beginning.

```
list_example[-1]  ────────▶  40  last element

list_example[-2]  ────────▶  30  second to last element
```

Table 4.3 illustrates how to perform sequence operations on a Python list.

Table 4.3.: Python list operations. list_example = [10, 20, 30, 40]

Operation	Example	Result
retrieving	list_example[2]	30
updating	list_example[1] = 50	[10, 50, 30, 40]
inserting	list_example.insert(2,50)	[10, 20, 50, 30, 40]
removing	del list_example[1]	[10,30,40]
appending	list_example.append(50)	[10, 20, 30, 40, 50]

The dot notation used in the list insert and list append operations is an example of object-oriented programming and will be explained more fully in Chapter 11. For now, think of the name to the right of the dot, called a method in object-oriented programming, as a function that applies to the object to the left of the dot.

The slice operation allows us to extract multiple list elements with one simple expression. The general form is a variation of the bracket notation, [m:n], which returns elements starting with the one indexed by m and extending up to, but not including, the element indexed by n. The second through the third elements of list_example could be extracted with list_example[1:3]. If the first index number, m, is omitted, then the slice begins with the first element of the list. If the last index number, n, is omitted, then the slice includes all elements through the last one. Figure 4.7 illustrates the use of the slice operation.

```
[2]  1 list1 = [10., 20., 30., 40., 50.]
     2 list2 = list1[1:3]
     3 list3 = list1[2:]
     4 list4 = list1[:3]
     5 print(list1)
     6 print(list2)
     7 print(list3)
     8 print(list4)

    [10.0, 20.0, 30.0, 40.0, 50.0]
    [20.0, 30.0]
    [30.0, 40.0, 50.0]
    [10.0, 20.0, 30.0]
```

Figure 4.7.: Examples of the slicing operation.

One more common operation frequently performed on a sequence, including Python lists, is traversal. Traversing a sequence involves accessing each element in order. A simple way to do this is with a for loop. For example, printing out each element of a list on a separate line we can be done as follows:

```
for e in list_example:
    print(e)
```

4.3. Python tuples

A **tuple** is an immutable Python sequence data type. Tuples are defined using parentheses ().

```
(10, 20, 30)    ('apple', 'banana', 'cherry')
```

To avoid ambiguity, a tuple with a single element must contain the final comma.

```
tuple_example = (10,)
```

In this example, if tuple_example did not contain the final comma, then the Python interpreter would assume it is an integer. This is shown in the Colab notebook segment below.

```
[1]   1 tuple_example = (10,)
      2 type(tuple_example)

   tuple
```

```
   1 tuple_example = (10)
   2 type(tuple_example)

   int
```

The index syntax applies to tuples, just as it did for lists.

```
tuple_example = ('apple', 'banana', 'cherry')

tuple_example[0] ────────▶  'apple'    first element

tuple_example[1] ────────▶  'banana'   second element

tuple_example[2] ────────▶  'cherry'   third element
```

Remember that a tuple is immutable, meaning it cannot be changed after it is defined. Attempting to change a tuple element will result in an error, as shown in the notebook output below.

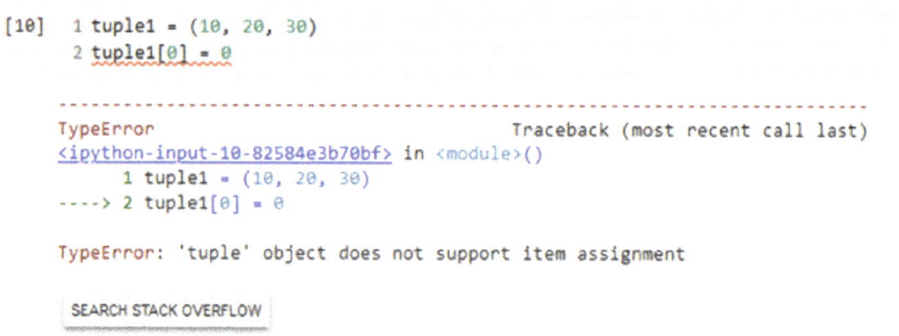

```
[10]   1 tuple1 = (10, 20, 30)
       2 tuple1[0] = 0
```

```
TypeError                              Traceback (most recent call last)
<ipython-input-10-82584e3b70bf> in <module>()
        1 tuple1 = (10, 20, 30)
----> 2 tuple1[0] = 0

TypeError: 'tuple' object does not support item assignment
```

SEARCH STACK OVERFLOW

Figure 4.8.: Error resulting from an attempt to assign to a tuple.

Python strings are an immutable sequence type, similar to tuples in behavior. The characters in a string are sequence elements, so they can be accessed using the bracket syntax.

```
string1 = 'FooBar'
string1[0]      ────────▶    'F' first element of the string
string1[3]      ────────▶    'B' fourth element of the string
```

Being immutable, any attempt to reassign a character in a string will result in an error.

4.4. NumPy Arrays

The NumPy library is a crucial building block for developing Python applications for scientific and engineering computing (*NumPy*, 2022). This library introduces an enhanced data structure for representing multidimensional arrays, methods of operating on these arrays, and many functions that can manipulate or extract information from these arrays. The NumPy library will be used in much of the code developed in the remainder of this book. Figure 4.9 illustrates the structure of arrays. A one-dimensional array is a type of sequence that behaves similarly to vectors in linear algebra. Two-dimensional arrays can be considered as matrices, and higher-dimensional arrays can be considered as tensors, as understood in certain areas of mathematics, such as differential geometry, engineering, and physics.

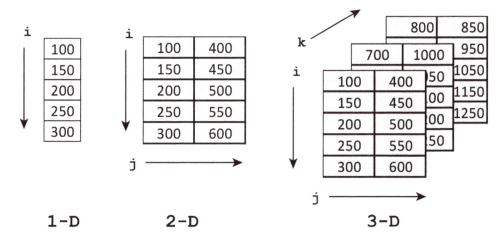

Figure 4.9.: Structure of 1, 2, and 3-dimensional arrays.

The array data structure or array object introduced by NumPy is called a ndarray, which stands for an n-dimensional array. The elements of a ndarray must all be of the same data type, called the dtype, such as integer or float. Two significant properties are associated with a specific ndarray: the array elements' shape and the data type (dtype). The shape of a ndarray is a tuple that specifies the size of the array along each dimension.

We will start with one-dimensional arrays to develop our understanding of NumPy arrays and operations on them. One way to create a NumPy array is to use the array routine in NumPy.

```
import numpy as np
a = np.array([10., 20., 30, 40.], dtype='float64')
```

Before using NumPy arrays and operations on them, we must first import the

NumPy package. Renaming the package 'np,' while unnecessary, has become traditional in data science applications. The first argument of the array routine is a Python list. The second argument specifies the data type of the elements in the array. The shape and data type of a NumPy array can be obtained as follows.

```
[2]   1 a = np.array([10.,20.,30., 40.], dtype='float64')
      2 a.shape

      (4,)
```

```
[3]   1 a.dtype

      dtype('float64')
```

Note that the shape is given as a tuple and shows that a is a one-dimensional array with four elements. The individual values in a NumPy array can be retrieved using the square bracket notation.

```
[4]   1 a[2]

      30.0
```

Now let us move on to a two-dimensional NumPy array.

```
[9]   1 a2d = np.array([[10.,20.,30., 40.], [50., 60., 70., 80.]],dtype='float64')
      2 a2d

      array([[10., 20., 30., 40.],
             [50., 60., 70., 80.]])
```

```
[10]  1 a2d.shape

      (2, 4)
```

The shape of a2d indicates that there are two dimensions. The first value in the shape tuple gives the number of rows, and the second value gives the number of columns. We can retrieve the value of individual elements in the array using square bracket notation. For example, the element in the second row and fourth column can be accessed as follows.

```
[11]   1 a2d[1,3]

       80.0
```

The first index number gives the row, and the second index number provides the column. Recall that Python uses a zero-based indexing system.

One advantage of using NumPy arrays instead of Python lists is that numerical calculations can be coded more succinctly. Suppose we have an array of numbers, and we need to add a constant value to each array member. This can be done with one line of code if the array is represented in a NumPy array.

```
a1d = np.array({[}10., 20., 30.{]})
b1d = a1d + 5.0
```

If a1d were a Python list, then the arithmetic expression would yield an error. We would need to set up a loop to add the constant to each element explicitly.

4.5. Basic Input and Output with Files

A basic operation that any data-oriented computational scientist must perform is reading columns of numerical data from a file into a program for processing and then writing out columns of numerical data to a file for use at a later time. In this section, we will learn about one method of performing these operations that uses a function in the NumPy module, one that will take care of many of the tedious processes that arise in performing reading and writing numerical data.

Let's work with a specific example. The file GLB.Ts+dSST.txt is a plain text file that contains the global mean temperature anomaly data compiled by NASA's Goddard Institute for Space Studies Surface Temperature Analysis project, shown in Table 4.1. Figure 4.10 shows a screenshot of this file viewed with a plain text editor. We see that the file has eight lines of comments that describe the dataset, one line that specifies column headings, and then 20 columns of numbers. Spaces separate the numbers in each row, so we call this space-delimited data.

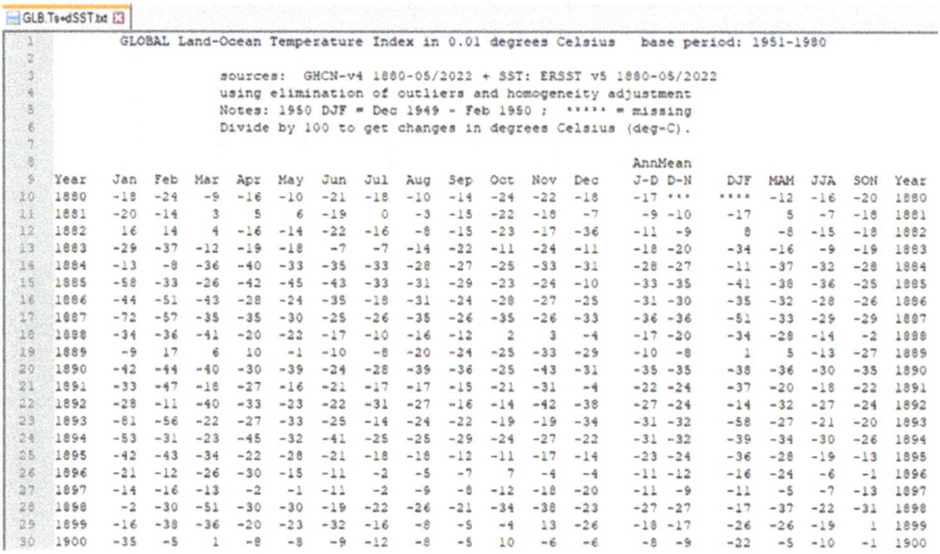

```
GLB.Ts+dSST.txt
        GLOBAL Land-Ocean Temperature Index in 0.01 degrees Celsius   base period: 1951-1980

                    sources:  GHCN-v4 1880-05/2022 + SST: ERSST v5 1880-05/2022
                    using elimination of outliers and homogeneity adjustment
                    Notes: 1950 DJF = Dec 1949 - Feb 1950 ;  ***** = missing
                    Divide by 100 to get changes in degrees Celsius (deg-C).

                                                                        AnnMean
Year   Jan  Feb  Mar  Apr  May  Jun  Jul  Aug  Sep  Oct  Nov  Dec    J-D D-N    DJF  MAM  JJA  SON  Year
1880   -18  -24   -9  -16  -10  -21  -18  -10  -14  -24  -22  -18    -17 ***   ****  -12  -16  -20  1880
1881   -20  -14    3    5    6  -19    0   -3  -15  -22  -18   -7     -9 -10    -17    5   -7  -18  1881
1882    16   14    4  -16  -14  -22  -16   -8  -15  -23  -17  -36    -11  -9      8   -8  -15  -18  1882
1883   -29  -37  -12  -19  -18   -7   -7  -14  -22  -11  -24  -11    -18 -20    -34  -16   -9  -19  1883
1884   -13   -8  -36  -40  -33  -35  -33  -28  -27  -25  -33  -31    -28 -27    -11  -37  -32  -28  1884
1885   -58  -33  -26  -42  -45  -43  -33  -31  -29  -23  -24  -10    -33 -35    -41  -38  -36  -25  1885
1886   -44  -51  -43  -28  -24  -35  -18  -31  -24  -28  -27  -25    -31 -30    -35  -32  -28  -26  1886
1887   -72  -57  -35  -35  -30  -25  -26  -35  -26  -35  -26  -33    -36 -36    -51  -33  -29  -29  1887
1888   -34  -36  -41  -20  -22  -17  -10  -16  -12    2    3   -4    -17 -20    -34  -28  -14   -2  1888
1889    -9   17    6   10   -1  -10   -8  -20  -24  -25  -33  -29    -10  -8      1    5  -13  -27  1889
1890   -42  -44  -40  -30  -39  -24  -28  -39  -36  -25  -43  -31    -35 -35    -38  -36  -30  -35  1890
1891   -33  -47  -18  -27  -16  -21  -17  -17  -15  -21  -31   -4    -22 -24    -37  -20  -18  -22  1891
1892   -28  -11  -40  -33  -23  -22  -31  -27  -16  -14  -42  -38    -27 -24    -14  -32  -27  -24  1892
1893   -81  -56  -22  -27  -33  -25  -14  -24  -22  -19  -19  -34    -31 -32    -58  -27  -21  -20  1893
1894   -53  -31  -23  -45  -32  -41  -25  -25  -29  -24  -27  -22    -31 -32    -39  -34  -30  -26  1894
1895   -42  -43  -34  -22  -28  -21  -18  -18  -12  -11  -17  -14    -23 -24    -36  -28  -19  -13  1895
1896   -21  -12  -26  -30  -15  -11   -2   -5   -7    7   -4   -4    -11 -12    -16  -24   -6   -1  1896
1897   -14  -16  -13   -2   -1  -11   -2   -9   -8  -12  -18  -20    -11  -9    -11   -5   -7  -13  1897
1898    -2  -30  -51  -30  -30  -19  -22  -26  -21  -34  -38  -23    -27 -27    -17  -37  -22  -31  1898
1899   -16  -38  -36  -20  -23  -32  -16   -8   -5   -4   13  -26    -18 -17    -26  -26  -19    1  1899
1900   -35   -5    1   -8   -8   -9  -12   -8   -5   10   -6   -6     -8  -9    -22   -5  -10   -1  1900
```

Figure 4.10.: Screenshot of the text file containing NASA temperature.

The NumPy package provides a function for importing numerical data from a text or CSV file as long as the data is in the form of columns of numbers. Since this is a common situation in scientific and engineering applications, we will use this technique here. The NumPy function we will use is loadtxt. While loadtxt can skip rows of non-numerical data at the beginning of the file, only numerical data must be present in the imported columns. The file shown in Figure 4.10 has missing numbers, indicated by ***, and the presence of these characters will cause an error. To use loadtxt, we must do some initial cleanup of the data file. One way to accomplish this cleanup is to open the file in a plain text editor and delete the rows that have missing numbers. The revised file is called GLB.Ts+dSST_clean.txt.

Keeping the original file unchanged is a good practice in case problems arise in the data processing workflow.

The code that will read in the data and separate the year column and the annual mean temperature anomaly column is shown below.

```python
import numpy as np

# Read in data and extract columns
cols = np.loadtxt('GLB.Ts+dSST_clean.txt', skiprows=9)
year = cols[:, 0]
annual_temp_anomaly = cols[:, 13]

# Calculate the actual annual temperature anomaly
annual_temp_anomaly = annual_temp_anomaly/100.
```

Figure 4.11.: Code for reading in columns of numerical data.

If you are using Google Colab notebooks as your development environment, you will need to upload the data file to your Google Drive. You will then need to mount the Google Drive and change the working directory to wherever you placed your file, typically in the same folder as the Colab notebook file. This can be accomplished by adding the following code at the beginning of the Colab notebook.

```python
from google.colab import drive
drive.mount(\textquotesingle/content/drive\textquotesingle)
\%cd /content/drive/MyDrive/yourWorkingDirectory
```

You should replace 'yourWorkingDirectory' with the path to the folder where your data file is located on your Google Drive.

The function savetxt from the NumPy module can be used to write columns of numerical data to a text file. The first argument gives the file name to which data is written. The second argument provides the 2d numpy array containing the data columns. A delimiter can be specified with the delimiter keyword. Figure 4.13 shows the output file. By default, the format of the numbers will be scientific notation with the full precision for floats. The following code snippet, Figure 4.12, shows how to construct a 2d NumPy array from columns of data and then write out the data to a text file, a portion of which is shown in Figure 4.13.

```
# Create 2d array from year and temperature 1d arrays
outArray = np.column_stack((year, annual_temp_anomaly))

# Save to text file
np.savetxt('annual_temp_anomaly.txt', outArray, delimiter='\t')
```

Figure 4.12.: Code for saving columns of numerical data to a text file.

annual_temp_anomaly.txt ⊠		
1	1.881000000000000000e+03	-8.999999999999999667e-02
2	1.882000000000000000e+03	-1.100000000000000006e-01
3	1.883000000000000000e+03	-1.799999999999999933e-01
4	1.884000000000000000e+03	-2.800000000000000266e-01
5	1.885000000000000000e+03	-3.300000000000000155e-01
6	1.886000000000000000e+03	-3.099999999999999978e-01
7	1.887000000000000000e+03	-3.599999999999999867e-01
8	1.888000000000000000e+03	-1.700000000000000122e-01
9	1.889000000000000000e+03	-1.000000000000000056e-01
10	1.890000000000000000e+03	-3.499999999999999778e-01
11	1.891000000000000000e+03	-2.200000000000000011e-01

Figure 4.13.: Part of the text output using the savetxt function.

We can specify the format for the numbers written out using the fmt parameter. This parameter is a string with the following form:

%[flag]width[.precision]specifier

where each part is described below:

flags (optional):
- : left justify

+ : Forces to precede result with + or -.

0 : Left pad the number with zeros instead of space (see width).

width: The minimum number of characters to be printed. The value is not truncated if it has more characters.

precision (optional):

For integer specifiers (e.g., d, i,o,x), the minimum number of digits.

For e, E, and f specifiers, the number of digits to print after the decimal point.

For g and G, the maximum number of significant digits.

For s, the maximum number of characters.

specifiers:

c: character

d or i: a signed decimal integer

e or E: scientific notation with e or E.

f: decimal floating point

g, G: use the shorter of e, E or f

o: signed octal

s: a string of characters

u: an unsigned decimal integer

x, X: unsigned hexadecimal integer

The fmt parameter can be a single format specifier, which will be applied to all the columns of numbers, or it can be a list of format specifiers, one for each column. Since the first column of the output is supposed to be a year, it would be more natural to format that number as an integer. This can be accomplished with the following statement.

```
# Save to text file
header_string = 'year\t temp anomaly'
np.savetxt('annual_temp_anomaly.txt', outArray, fmt=['%6i','%10.2e'],
           delimiter='\t', header=header_string)
```

A header that gives column labels is provided with the header keyword argument.

A portion of the output file is shown in Figure 4.14.

```
 annual_temp_anomaly.txt ☒
  1   # year       temp anomaly
  2     1881       -9.00e-02
  3     1882       -1.10e-01
  4     1883       -1.80e-01
  5     1884       -2.80e-01
  6     1885       -3.30e-01
  7     1886       -3.10e-01
  8     1887       -3.60e-01
  9     1888       -1.70e-01
 10     1889       -1.00e-01
 11     1890       -3.50e-01
```

Figure 4.14.: A portion of savetxt output with formatted numbers.

4.6. Introduction to Scientific Visualization with Matplotlib

We are ready to create our first scientific visualization: a two-dimensional scatter graph showing the relationship between the mean global temperature anomaly and year. There are several standard plotting packages available in Python, but we will focus on Matplotlib since it is widely used and forms the basis for other visualization packages (The Matplotlib Development team, 2021). We will discuss scientific visualizations in greater detail in Chapter 5. Here, we want to touch on the basics of using the Matplotlib package to create a 2d scatter graph.

Let us assume that the mean global surface temperature anomaly values and corresponding year are in the data file annual_temp_anomaly.txt. The following code produces a simple scatter graph of this data.

```python
import numpy as np
import matplotlib.pyplot as plt

# Read in data
cols = np.loadtxt('annual_temp_anomaly.txt',skiprows=1)
```

```
year_data = cols[:,0]
temp_anomaly_data = cols[:,1]

# Plot data
plt.plot(year_data, temp_anomaly_data, linestyle='', marker='d', markersize
    =5.0)
plt.xlabel('year')
plt.ylabel('Temperature Anomaly')
plt.savefig('tempAnomalyVsYear.png', dpi=300)
plt.show()
```

The pyplot sub-package of the Matplotlib package is imported and renamed plt.
This is traditional in data science applications. The entire Matplotlib package is
not needed to create most of the plots we use. The plot is created with the Matplotlib plot function. The first argument is the 1d array containing the x-axis values.
The second argument is the 1d array with the y-axis values. The other keyword
arguments are set, so only data points are graphed with a diamond data marker.
The markersize is always adjusted to achieve a good visualization. Axis labels are
added with the xlabel and ylabel functions.

Finally, the plot is saved as a graphics file with the savefig function. The graphics
file format is determined from the extension on the chosen file name, given as the
first argument. The dpi keyword sets the resolution. The show function is not
needed if the code is executed in a Jupyter notebook but will be required if it is
executed in other programming environments. The plot is shown in Figure 4.15.

We will discuss additional methods of customizing plots in later chapters.

4.7. Computational Problem Solution: Identifying Patterns in Global Surface Temperature Data

We will now return to the problem introduced at the beginning of the chapter: how
to use data visualization to help identify temperature anomaly trends over time.

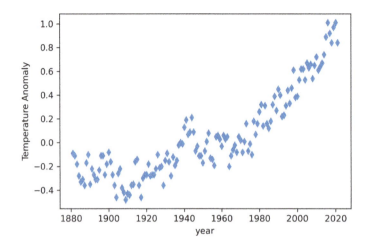

Figure 4.15.: Temperature anomaly as a function of year produced by Matplotlib.

4.7.1. Analysis

We start with a precise description of the problem and some features of the solution that will allow us to recognize a successful solution. The problem is to start with the data file given by NASA, GLB.Ts+dSST.txt (GISTEMP Team, 2022a), and produce a graph, similar to Figure 4.1, that performs some kind of smoothing of the widely fluctuating temperature anomaly values. We can recognize a solution when the graph has a smoothing curve displayed that shows fewer yearly fluctuations compared to the annual raw data.

Some quite sophisticated statistical methods exist for smoothing random fluctuating time series data, including the LOWESS smoothing shown in Figure 4.1. Using such techniques requires considerably more mathematical sophistication than we want to use here. Therefore, we will use a simple moving average method that is conceptually simple and relatively easy to implement computationally.

Consider a subset of the annual mean temperature anomaly data from Figure 4.14, shown in Table 4.4.

Table 4.4.: A subset of temperature anomaly data.

year	temperature anomaly [C]
1883	-0.18
1884	-0.28
1885	-0.33
1886	-0.31
1887	-0.36

The 5-year moving average for the 1885 temperature anomaly is calculated from

$$T_m\,(1885) = \frac{(-0.18 - 0.28 - 0.33 - 0.31 - 0.36)}{5.0}\,[C] = -0.29\,[C] \qquad (4.1)$$

We would replace the 1885 temperature anomaly in the data set with this new value. The moving average for each year is calculated similarly. We move the 5-year window, two years before the chosen year, two years after the selected year, plus the chosen year, to each year in turn. If $T\,(y)$ represents the mean temperature anomaly for year y, then the general formula for the 5-year moving average is

$$T_m\,(y) = \frac{T\,(y-2) + T\,(y-1) + T\,(y) + T\,(y+1) + T\,(y+2)}{5} \qquad (4.2)$$

We could try different window sizes, 7-year, 9-year, and so on, to find one that clarifies a trend, but we will stick with the 5-year window for this solution.

4.7.2. Design

The workflow for our problem solution is

1. Clean up the raw NASA data file GLB.Ts+dSST.txt so that it does not contain non-numerical data, except for the header comments.
2. Note the first year for which a 5-year moving average can be calculated (defines first_year).
3. Note the last year for which a 5-year moving average can be calculated (defines last_year).

4. Write code to import the year and annual mean columns from the clean data file.
5. Convert the annual mean column numbers to Celsius by dividing them by 100.
6. Write code that can calculate the 5-year moving average temperature anomaly for a general specified year, named current_year.

Step 1 is performed with a basic text editor application, and we assume that the cleaned-up data file is GLB.Ts+dSST_clean.txt.

For the NASA data set we are using here, the first year for which a 5-year moving average can be calculated is 1883, and the last year is 2019.

Steps 4 and 5 were discussed in section 4.5. These steps define the year_data and annual_temp_anomaly arrays.

Step 6 requires some thought. Assume that year_index is the index number giving the year currently being considered. To expedite using the Matplotlib plot function, we will also define two new 1d NumPy arrays: moving_year_data will contain all the years for which we calculate a moving average, and moving_annual_temp_anomaly will contain the corresponding calculated moving averages. A pseudocode snippet of this process is shown below.

```
Extract the subset of temperature anomalies from annual_temp_anomaly
  (this defines anomaly_subset)
Calculate the mean of the numbers in anomaly_subset (defines
  temp_anomaly_mean)
Add the temp_anomaly_mean to moving_annual_temp_anomaly
```

All of the variables required for the problem solution are listed in Table 4.5.

4.7.3. Implementation

We can collect together pieces of code from sections 4.5 and 4.6 to create code for our problem solution. Since NumPy arrays do not allow for the append operation,

Table 4.5.: List of data structures required for the problem solution.

Data Structure	Type	Description
first_year	float variable	first year for the 5-year moving average calculation
last_year	float variable	last year for the 5-year moving average calculation
year_data	1-d numpy array	contains the column of year values in the NASA data set
annual_temp_anomaly	1-d numpy array	contains the column of annual temperature anomaly values in the NASA data set
year_index	integer variable	specifies the year currently being calculated
anomaly_subset	1-d numpy array	contains the five temperature anomalies that will be averaged
moving_year_data	1-d numpy array	contains the year values for which a moving average is calculated
moving_annual_temp_anomaly	1-d numpy array	contains the moving average value for each year in moving_year_data

as can be done with a Python list, we must create new NumPy arrays with the correct length before using them. An empty array is created using the NumPy empty function, as shown below in lines 29 and 30.

The subset of the temperature anomaly values to be averaged is obtained from an array slicing operation in line 36. The average of the subset values is obtained using the NumPy mean function in line 37.

```
1   """
2   Title: 5-year Moving Average Plot
3   Author: C.D. Wentworth
4   Version: 6.29.2022.1
5   Summary: This program will read in annual temperature anomaly data
6            provided by NASA, calculate a 5-year
7            moving average, then plot both the raw data and moving average.
8   Revision History:
9       6.29.2022.1: base
10
11  """
12  import numpy as np
13  import matplotlib.pyplot as plt
14
15  cols = np.loadtxt('GLB.Ts+dSST_clean.txt', skiprows=9)
16  year_data = cols[:, 0]
17  annual_temp_anomaly = cols[:, 13]
18
19  # Calculate the actual annual temperature anomaly
20  annual_temp_anomaly = annual_temp_anomaly/100.
21
22  # Define the year range
23  first_year = 1883.
24  last_year = 2019.
25  first_year_index = np.where(year_data == first_year)[0][0]
26  last_year_index = np.where(year_data == last_year)[0][0]
27
28  # Create empty numpy arrays for plot
29  moving_year_data = np.empty(len(year_data)-4)
30  moving_annual_temp_anomaly = np.empty(len(moving_year_data))
31
32  # Calculate the moving average
33  moving_year_index = 0
34  for year_index in range(first_year_index, last_year_index + 1):
35      # obtain temperature anomaly subset
36      anomaly_subset = annual_temp_anomaly[year_index - 2:year_index + 2]
37      temp_anomaly_mean = np.mean(anomaly_subset)
38      moving_year_data[moving_year_index] = year_data[year_index]
39      moving_annual_temp_anomaly[moving_year_index] = temp_anomaly_mean
40      moving_year_index = moving_year_index + 1
```

The code block that creates the plot is shown below. The linestyle keyword argument is used to choose whether a line is solid, dotted, or dashed. The empty string, '', indicates no line is drawn, which is often used when plotting data points. The linestyle value '-' indicates a solid line. The label keyword argument specifies a string that is used when a legend is shown. The legend is produced with default placement by

```
plt.legend()
```

We will discuss keyword arguments used in the plot function in greater detail in the next chapter. Generally, a review of examples from the Matplotlib documentation or tutorials can help figure out good choices for the keyword argument values (The Matplotlib Development Team, 2022).

```
42  # plot the raw data and moving average
43  plt.plot(year_data, annual_temp_anomaly, linestyle='', marker='d',
44          markersize=5.0, label='Annual Mean')
45  plt.plot(moving_year_data, moving_annual_temp_anomaly, linestyle='-',
46          linewidth=2, label='5 year Moving Average')
47  plt.xlabel('year')
48  plt.ylabel('Temperature Anomaly')
49  plt.legend()
50  plt.savefig('tempAnomalyVsYearMovingAvg.png', dpi=300)
51  plt.show()
```

4.7.4. Testing

When the 5-year Moving Average Plot program is executed, we get the graph shown in Figure 4.16. This plot does resemble the one produced by NASA scientists, shown in Figure 4.1. This provides evidence that the problem solution is correct. The five year moving average line shows more fluctuations than the LOWESS smoothing line in Figure 4.1. This is likely due to the five year moving window being smaller than that used in the LOWESS smoothing algorithm. It would be interesting to compare a seven year moving average with the five year moving average.

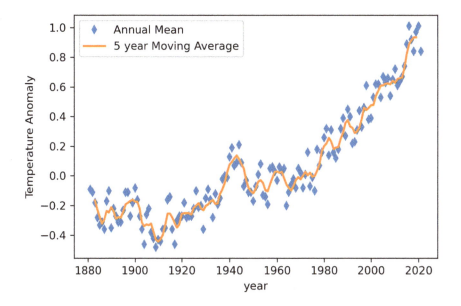

Figure 4.16.: Plot output from the 5-year Moving Average Plot program.

4.8. Exercises

1. (Fill-in-the-blank): A **sequence** is a _____ ordered set of values, each of which can be accessed using an _____ number

2. Which of the following Python data structures are mutable? Choose all that apply.

a. lists

b. strings

c. tuples

d. NumPy arrays

3. What is the resulting list when the value at index three is removed?

index	value
0	25
1	15
2	-5
3	50
4	75

index	value
0	25
1	15
2	-5
3	0
4	75

(a)

index	value
0	25
1	15
2	-5

(b)

index	value
0	25
1	15
2	-5
4	75

(c)

index	value
0	25
1	15
2	-5
3	0
4	0

(d)

4. What are the index values for a tuple with 4 elements?

a. 0-4

b. 0-3

c. 1-4

5. How would you define a tuple with the one element 12?

a. [12]

b. [12,]

c. (12)

d. (12,)

6. What would be the list resulting from the [1:3] slice in

[80, 70, 75, 50, 65]

  a. [80, 70, 75]

  b. [70, 75, 50]

  c. [70, 75]

7. Select all true statements about Python sequence data types.

  a. All items in a Python list must be the same data type (all integers, all strings, etc.).

  b. An attempt to assign a new value to a tuple element will result in an error.

  c. Python lists, tuples, and strings all use zero-based indexing.

8. Write a Python expression that prints out the third element of list1 defined below.

list1 = [125, -75, 25, 50, 15]

4.9. Program Modification Problems

1. In this exercise, you are given a program, shown below, that generates a list of random numbers and then counts the number of odd numbers in the list. You need to modify the program so that it creates two lists of random numbers, adds up the odd numbers in one list, adds up the even numbers in the second list, and then adds the two sums together.

```
"""
Program: Count Odds
Author: C.D. Wentworth
Version: 2.5.2020.1
Summary: This program creates a list of random numbers and
         counts how many odd numbers are in the list.

"""
import random as rn

# make a list of random numbers
lst = []
for i in range(10):
    lst.append(rn.randint(0, 1000))

# count the number of odd numbers in the list
odd = 0
for e in lst:
    if e % 2 != 0:
        odd = odd + 1
print('The number of odd numbers in the list is: ',odd)
```

2. The program below will create a tan(t) graph. Modify it to create a plot of the following function

$$f(t) = 5.0 * \sin(t)$$

for $-3.14 \le t \le 3.14$.

The plot should have

- a dashed red line with a thickness of 5.
- The x-axis title should be 't.'
- The y-axis title should be 'f(t).'
- The y-axis scale should be adjusted to $-6 \le t \le 6$
- The chart title should be '5*sin(t) versus t'. The font size should be 20. The font color should be green.

```
"""
```

```
Program: Plot Function
Author:
Version: 1.25.2020.1
Summary: This script creates a basic plot of a function with
         user chosen features including the line style and
         line width.

"""

import matplotlib.pyplot as plt
import numpy as np
x = np.linspace(-3,3,100)
y = np.tan(x)
plt.plot(x,y,color='g',linestyle='-',linewidth=4)
plt.xlabel('x',fontsize=16)
plt.ylabel('y',fontsize=16)
plt.title('tan(x) vs x',fontsize=24, color='red')
plt.grid(True)
plt.axis(ymin=-10,ymax=10)
plt.show()
```

3. Write a program that creates a plot of $\cos(t)$ and $\sin(t)$ for $0 \leq t \leq 2\pi$ with both plots on the same graph. You can start with the Chapter 4 Program Modification Problem 2 code. Your final graph should

- show sin as a solid red curve
- show cos as a solid blue curve
- have a legend
- show the title 'Comparison of cos and sin' in green.
- use $-1.5 \leq y \leq 1.5$ for the y-axis scale

4.10. Program Development Problems

1. Write a program that will create a moving average plot of the NASA global temperature anomaly data for a user specified moving average window. The user will specify a window size that should be an odd number of years. The program should calculate the required moving average for as many years in the data as possible and then create a plot of the annual data, shown with data symbols, and

the moving average, shown as a solid line.

4.11. References

GISTEMP Team. (2022a). *Data.GISS: GISS Surface Temperature Analysis (v4): Analysis Graphs and Plots.* https://data.giss.nasa.gov/gistemp/graphs/

GISTEMP Team. (2022b). *GISS Surface Temperature Analysis (GISTEMP), version 4.* [Database]. NASA Goddard Institute for Space Studies. https://data.giss.nasa.gov/gistemp/

NumPy. (2022). https://numpy.org/

The Matplotlib Development team. (2021). *Matplotlib—Visualization with Python.* https://matplotlib.org/

The Matplotlib Development Team. (2022). *Tutorials—Matplotlib 3.5.2 documentation.* https://matplotlib.org/stable/tutorials/index.html

5. Principles of Scientific Visualization

Motivating Problem: Comparing CO₂ Emissions by Country

Carbon dioxide is a significant greenhouse gas, and fossil fuel use is the major contributor to its increase in the earth's atmosphere. Understanding the emissions from fossil fuel use by different countries is an integral part of public policy discussions. The US Department of Energy maintains a database of such data that can be used by environmental scientists, economists, and other professionals (Boden et al., 2013). A subset of the data is shown below in Table 5.1. The entire dataset is available in the file fossil-fuel-co2-emissions-by-nation.csv.

Table 5.1.: CO_2 emission from fossil fuel use

Year	Country	Total	Solid Fuel	Liquid Fuel	Gas Fuel	Cement	Gas Flaring	Per Capita	Bunker fuels (Not in Total)
1800	CANADA	1	1	0	0	0	0	0	0
1800	GERMANY	217	217	0	0	0	0	0	0
1800	POLAND	111	111	0	0	0	0	0	0
1800	UNITED KINGDOM	7269	7269	0	0	0	0	0	0
1800	UNITED STATES OF AMERICA	69	69	0	0	0	0	0	0
1801	CANADA	1	1	0	0	0	0	0	0
1801	GERMANY	146	146	0	0	0	0	0	0
1801	POLAND	121	121	0	0	0	0	0	0
1801	UNITED KINGDOM	7290	7290	0	0	0	0	0	0
1801	UNITED STATES OF AMERICA	73	73	0	0	0	0	0	0
1802	CANADA	1	1	0	0	0	0	0	0
1802	FRANCE (INCLUDING MONACO)	611	611	0	0	0	0	0	0
1802	GERMANY	151	151	0	0	0	0	0	0
1802	POLAND	123	123	0	0	0	0	0	0
1802	UNITED KINGDOM	7328	7328	0	0	0	0	0	0
1802	UNITED STATES OF AMERICA	79	79	0	0	0	0	0	0

A complete description of the data in each column is provided in Table 5.2.

The computational problem we want to solve in this chapter is to develop visualizations of the data in this table that can help in discussing the issue of greenhouse

Table 5.2.: Column description for data in Table 5.1.

Field Name	Order	Type (Format)	Description
Year	1	year	Year
Country	2	string	Nation
Total	3	number	Total carbon emissions from fossil fuel consumption and cement production (million metric tons of C)
Solid Fuel	4	number	Carbon emissions from solid fuel consumption
Liquid Fuel	5	number	Carbon emissions from liquid fuel consumption
Gas Fuel	6	number	Carbon emissions from gas fuel consumption
Cement	7	number	Carbon emissions from cement production
Gas Flaring	8	number	Carbon emissions from gas flaring
Per Capita	9	number	Per capita carbon emissions (metric tons of carbon; after 1949 only)
Bunker fuels (Not in Total)	10	number	Carbon emissions from bunker fuels (not included in total)

gas emissions from fossil fuel use.

5.1. What are Scientific Visualizations?

Everyone knows the quaint aphorism "A picture is worth 1000 words". It communicates that a visual representation of information can often inform more quickly and accurately than a simple verbal description. When scientists and engineers wish to understand large amounts of data or interpret the results of a model, the same principle applies: visual representations are often more useful than the data viewed in its original form. In the age of Big Data and computer simulations, this observation takes on even more meaning and leads to the field of scientific visualization.

An excellent example of this bit of wisdom is seen in comparing Table 4.1 and Figure 4.1. The graph of temperature anomaly as a function of year identifies patterns much more straightforwardly than just looking at the data table itself. Similarly, we want to develop visualizations of the data in Table 5.1 that will help us identify useful patterns or trends that are difficult to see just by looking at the columns of numbers. Of course, identifying structure using visualization techniques requires some thought about how to create visualizations, and that is the subject of the scientific visualization field.

5.1.1. Definition of Scientific Visualization

As you might guess, allowing a bunch of academics to define a subject will lead to as many definitions as there are academics. But we can distill some common features that get us oriented to the topic. Here is our working definition of scientific visualization (Ausoni et al., 2014):

Scientific visualization is concerned with graphically representing scientific phenomena to gain understanding and insight into the system that was previously impossible.

■ This may be part of the research process: graphics are used for understanding, interpretation, and exploration and may guide the direction of the research itself, from tweaking parameters to raising new questions.

■ It may be used in production environments, such as medical procedures, as part of a larger mission.

■ It may be used for educational purposes, in the classroom, etc.

5.2. Classifying Scientific Visualizations

There are many approaches to classifying scientific visualizations, and our goal here is not to perform an exhaustive review. Indeed, there is a rich literature on classifying data visualizations. Some focus on the type of data being visualized (Shneiderman, 2003); others focus on the models and algorithms used with data rather than the data itself (Tory & Moller, 2004). Instead, we will focus on one basic model that should be useful for novice computational scientists to develop effective visualizations. We will classify visualizations according to three principal axes, as shown in Figure 5.1: the content type of visual model (data-based - conceptual), dimensionality of the representation (2D-3D), and the element of time in the visualization (static - dynamic).

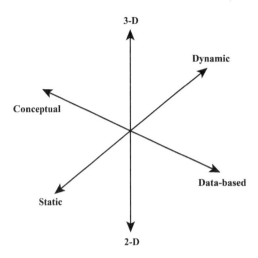

Figure 5.1.: Types of visualization models.

Figure 5.2 shows an example of a conceptual visualization. It illustrates the main elements of the earth's energy budget. While there is some quantitative information in the visualization, the primary purpose and information content are conceptual. Figure 5.3 shows the measured insolation (incoming solar radiation) as a function of the month for Lincoln, Nebraska (Power Project Team, n.d.). This is an example of a data-based visualization. It is easy to pick out the seasonal pattern when viewing the data in this form. Figure 5.3 also serves as an example of a 2D visualization. There are many types of 2D graphs, including XY, contour, and bar. Figure 5.3 is a bar graph. 2D graphs are the bread and butter for traditional scientific publications, so we will spend quite a bit of time learning how to produce these.

Both Figure 5.2 and Figure 5.3 are considered static visualizations since they are both static images that do not change. Dynamic visualizations such as animations that change over time are also very useful for exploring scientific data, particularly when we are interested in looking at changes over time. The following link shows the monthly average solar radiation arriving at earth's surface for one year: **Net Radiation Animation** (Shinker, 2016). Figure 5.4 shows three frames from the animation. This visualization is a dynamic, data-based, 2D visualization. Dynamic visualizations require more sophisticated viewing technologies than standard publications.

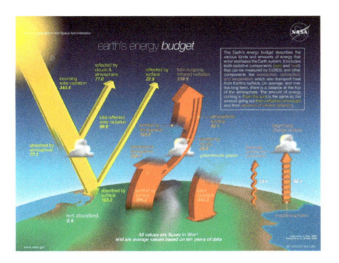

Figure 5.2.: Earth's energy budget. (Atkinson, 2017)

5.3. Python Modules for Visualization

There are several modules that make up a useful Python visualization environment. These include Numpy, Matplotlib, Pandas, and Seaborn. We introduced numpy and matplotlib previously. In this section we will cover some additional features of Matplotlib and then introduce the Pandas and Seaborn modules.

5.3.1. Creating 2d Graphs Using Matplotlib

In Chapter 4 we saw how to create a basic scatter graph using the plot function in the Matplotlib.pyplot module. We will now learn to customize a 2D plot. We will start with plotting functions. The following code illustrates specifying line color and line style. Figure 5.5 shows the result of this code.

```python
import matplotlib.pylab as plt
import numpy as np

x = np.linspace(0,5,40)
y1 = 0.10*np.exp(0.10*x)
y2 = 0.10*np.exp(0.20*x)
y3 = 0.10*np.exp(0.30*x)
plt.plot(x,y1,color='k' , label='mu= 0.1' , linestyle='-' , linewidth=3)
plt.plot(x,y2,color='b' ,label='mu= 0.2' ,linestyle='--', linewidth=3)
plt.plot(x,y3,color='r' ,label=' mu= .3' , linestyle=-.' , linewidth=3)
plt.xlabel('x' , fontsize=14)
plt.ylabel('y' , fontsize=14)
plt.legend()
```

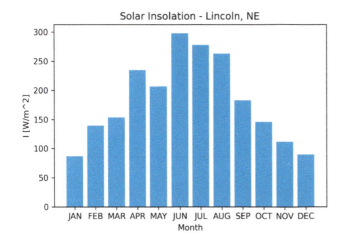

Figure 5.3.: Insolation (incoming solar radiation) as a function of month measured at the earth's surface in Lincoln, Nebraska.

There are several things to point out about the code.

■ The Numpy linspace function is used to generate a set of equally spaced numbers that can be used for the function calculations. This function generates a set of equally spaced numbers and returns them as a 1D numpy array. The syntax is

```
np.linspace(start, stop, num)
```

start: starting number of the interval

stop: ending number of the interval

num: the number samples to generate

■ The color of a line (or data symbol) is determined by the color keyword argument. Table 5.3 gives the codes for the base set of colors in Matplotlib. The CSS color list gives considerably more choice. See the Matplotlib documentation for the CSS color codes (*List of Named Colors — Matplotlib 3.5.2 Documentation*, 2022).

■ The thickness of the line is set by the linewidth keyword argument. The number can be any positive float.

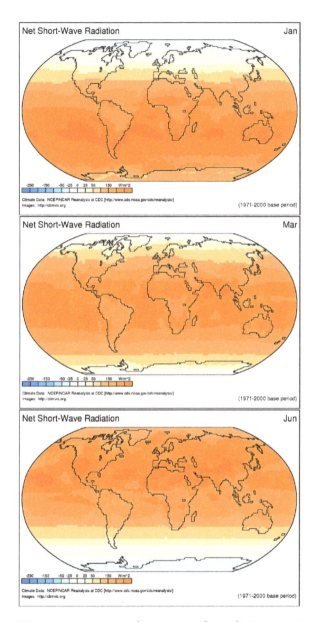

Figure 5.4.: Average short wave solar radiation arriving at Earth's surface.

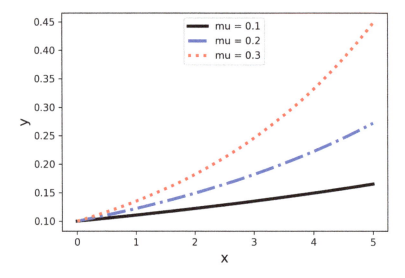

Figure 5.5.: Example of changing colors and linestyles.

The type of line drawn is determined by the linestyle keyword argument. The allowed values are listed in Table 5.4.

Table 5.3.: Matplotlib base colors.

description	color	code
blue		b
green		g
red		r
cyan		c
magenta		m
yellow		y
black		k
white		w

Table 5.4.: Matplotlib linestyle codes.

linestyle description	code	short code
Solid	'solid'	'-'
Dashed	'dashed'	'--'
Dotted	'dotted'	':'
Dashdot	'dashdot'	'-.'
None	'none'	''

If you do not like the default location of the legend, then you can customize it by using the loc keyword argument. The allowed strings or corresponding numerical codes are shown in Table 5.5.

Table 5.5.: Matplotlib legend loc values.

string	number code
best'	0
upper right'	1
upper left'	2
lower left'	3
lower right'	4
right'	5
center left'	6
center right'	7
lower center'	8
upper center'	9
center'	10

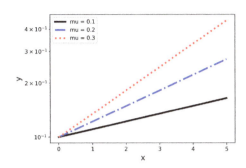

Figure 5.6.: Example of using the yscale function.

```
plt.legend(loc='upper center')
```

When numbers to be plotted cover a large order of magnitude range, or when we want to easily identify power law or exponential law behavior, using a logarithmic scale can be helpful. To make the y-axis use a logarithmic scale we can use the yscale function.

```
plt.yscale('log')
```

Adding this to the code that produced Figure 5.5 gives us Figure 5.6.

5.3.2. Introduction to the Pandas Module

The Pandas module is a basic component of the Python computational science environment and is particularly helpful for problems involving data analytics (The pandas development team, 2022). It provides a useful data structure for containing data, the dataframe, and many functions for working with a dataframe. To illustrate some of the features of Pandas we will use a classic data set used for educational purposes: the Fisher Iris Flower Data Set (Fisher, 1936). This data set can be obtained in a convenient electronic form as a csv file from Kaggle (*Iris Species*, n.d.). This data is in the file Iris.csv.

A dataframe is a two-dimensional data structure, visualized as a 2-D grid, where each column can contain a different datatype. The data does not need to be numerical, as in the case of NumPy arrays. Each column in the dataframe has a label that will appear above the column in the first row of the grid. Each row of the grid, except for the first row containing the column labels, is indexed by a row number starting with 0. Figure 5.7 is a Colab Notebook excerpt that illustrates how to read in data from a csv file to create a Pandas dataframe. The Pandas function read_-csv can be used to read in tab-delimited data, too, by using the header keyword argument (*Pandas.Read_csv — Pandas 1.4.3 Documentation*, n.d.).

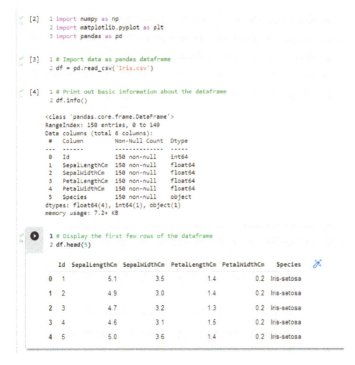

Figure 5.7.: Reading data into a dataframe and obtaining basic information about it.

An overview of the dataframe contents can be obtained by using the Pandas info function, as shown in Figure 5.7. We can see a listing of the column labels and how many non-null entries each column has. This last piece of information can be helpful in deciding what kind of data clean-up must be performed. The first few rows from the dataframe can be viewed by using the head function.

The data in an individual column from the dataframe can be accessed by using a square bracket notation, where the index in the bracket is the desired column label:

```
ds = df['Species']
```

In this example, the variable ds will be a pandas data series, which is a special pandas data structure that can be manipulated with appropriate pandas functions. Instead of working with pandas data series we will usually convert columns to Numpy arrays when the data type is numerical. This can be done with the to_numpy() method applied to a pandas data series. Here is an example:

```
SepalLength_data =
df['SepalLengthCm']}.to_numpy()
```

The variable SepalLength_data will be a 1D numpy array that can be used by matplotlib plotting functions, for example. Pandas has many plotting functions that can be used with dataframes, but we will usually convert dataframe columns to numpy arrays and use matplotlib for our plotting.

5.3.3. Introduction to the Seaborn Module

The Seaborn module is built on top of Matplotlib but offers some advantages in certain situations. Some of the advantages are

■ Integration with Pandas so that dataframes can be used directly without needing to convert columns to Numpy arrays.

■ Some useful graph types are prebuilt so that they can be implemented with a single function. One example is the pairwise or all-against-all scatterplot that we will use in 5.4.3.

■ There are some predefined styles that create aesthetically pleasing graphs consistent with good graphic design principles.

To illustrate some basic features of the Seaborn module we will use a selection of data from the NASA Exoplanet Archive (*NASA Exoplanet Archive*, n.d.). The data we will use is in the file PSCompPars_subset.csv. Before using the Seaborn module you must import it into your code file:

```
import seaborn as sns
import matplotlib.pyplot as plt
import numpy as np
import pandas as pd
```

Seaborn contains functions to create a wide variety of plots. One method of generating a scatterplot is to use the lmplot() function. The basic syntax is

```
fig = sns.lmplot(x=x_string, y=y_string, data=dataframe_name,
                fit_reg=False)
```

x_string: the x-axis column heading from the dataframe

y_string: the y-axis column heading from the dataframe

dataframe_name: the Pandas dataframe containing the data

If the fit_reg keyword argument is not set to False then a linear regression line is automatically added to the graph. We will come back to performing a linear regression fit to data in Chapter 10.

Here is an example of creating a scatterplot from the exoplanet data file.

```
planets =
pd.read_csv('PSCompPars_subset.csv', header=16)
planets_fig = sns.lmplot(x='pl_orbsmax', y='pl_orbper', data=planets,
    fit_reg=False)
                        plt.xlim(0,1.5)
plt.xlabel('Orbital Semi-major Axis [au]')
plt.ylabel('Orbital Period [days]')
planets_fig.savefig('pl_orbsmax-pl_orbper.png' , dpi=300)
```

Note that we applied three matplotlib.pyplot functions to the Seaborn plot, xlim, xlabel, and ylabel. This illustrates a nice feature of Seaborn. We can use matplotlib to provide some customization of the plot. This can be done because Seaborn is built on top of Matplotlib. Figure 5.8a shows the result of this code.

Overall style of a graph is determined by using one of the predefined themes. This

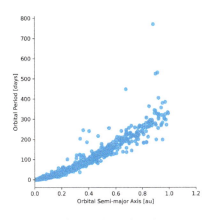

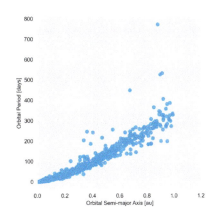

(a) Example of the Seaborn white theme.

(b) Example of the Seaborn darkgrid theme.

Figure 5.8.: Seaborn scatterplots.

is done with

```
sns.set_style(theme)
```

where theme is one of the following strings:

'darkgrid', 'whitegrid', 'dark', 'white', 'ticks'

Using the darkgrid theme to create the scatterplot shown in Figure 5.8a yields Figure 5.8b.

5.4. Creating a Good Visualization

Solving complex scientific and engineering problems can be facilitated by using a structured approach or strategy and the same is true for developing a good scientific visualization. We will use a strategy or workflow suggested by Ben Fry (Fry, 2008).

5.4.1. First Steps

The first step is to start with a question that you want to answer with the visualization. Starting with data and asking what it can tell us can lead to being overwhelmed by possibilities, whereas beginning with a question before even looking for or looking at data allows us to focus the visualization that we develop in a constructive way. A second part of this initial step is to define the audience for the visualization. Some helpful questions to define the audience include

Are they science professionals used to looking at graphs?

What information would they expect in a visualization?

How much time will they devote to looking at the visualization?

5.4.2. Visualization Development Workflow

Once we have defined a question and our audience, Fry suggests the following workflow (Fry, 2008):

1. Acquire - Obtain the data, whether from a file on a disk or a source over a network.
2. Parse or Understand the Data - Provide structure for the data's meaning, and order it into categories.
3. Filter – Clean the data and remove all but the data of interest.
4. Mine – Use methods from statistics, data mining, or more fundamental scientific principles to discern patterns or place the data in a mathematical context.
5. Represent - Choose a basic visual model, such as a bar graph, scatter graph, contour plot, or other visual construct.
6. Refine - Improve the basic representation to make it clearer and more visually engaging.
7. Interact - Add methods for manipulating the data or controlling what features are visible.

The last step, adding interaction, will not be discussed in this book.

Figure 5.9.: An iris flower showing how petal and sepal measurements are defined (Rathod, 2020).

5.4.3. Iris Flower Example

As an example, we will explore some properties of the iris flower contained in the classic educational data set originally created by R.A. Fisher (Fisher, 1936). The question guiding our visualization development is

> Can sepal or petal measurements be used to distinguish species of the iris flower?

The sepal is the part of the flower that protects the petals in the flower bud. Figure 5.9 shows an example of an iris flower.

We will step through the workflow defined above to develop visualizations that can help answer the question.

Acquire

We can easily **acquire** a nice dataset related to the iris flower already in a convenient electronic form as a csv file from Kaggle (*Iris Species*, n.d.). This data is in the

127

Table 5.6.: Description of the iris flower data set.

Number of Observations	150
Number of Attributes for each observation	5
Attributes observed	sepal length sepal width petal length petal width species
Number of null values	0

file Iris.csv.

Parse

We must **parse** or inspect the dataset to understand the properties or measurements that are included and whether any cleanup must be done. The pandas module contains some useful functions for inspecting a data frame. Therefore, we should first read the data in the Iris.csv file as a pandas data frame, as discussed in section 5.3. Next, we inspect the dataframe using the pandas info function. The output is shown in Figure 5.7. The output shows that all the columns are numerical data, except for the Species column. We can also see that there are 150 rows, or records, of data and that there are no null values. This will help us in the next step of filtering the data.

Filter

Filtering data involves removing data that will not be needed. Part of this process is to clean up the data by removing null values or removing records (rows in a pandas dataframe) that contain null values. We can determine whether there are null values by looking at the results from the pandas info function or by using the isnull function. The output from the info function applied to the iris flower dataframe already tells us that we do not have null values in the data set. The output from the isnull function, shown in Figure 5.10, can also tell us this result.

```
1 pd.isnull(df)
```

	Id	SepalLengthCm	SepalWidthCm	PetalLengthCm	PetalWidthCm	Species
0	False	False	False	False	False	False
1	False	False	False	False	False	False
2	False	False	False	False	False	False
3	False	False	False	False	False	False
4	False	False	False	False	False	False
...	...	...	...	...	...	...
145	False	False	False	False	False	False
146	False	False	False	False	False	False
147	False	False	False	False	False	False
148	False	False	False	False	False	False
149	False	False	False	False	False	False

150 rows × 6 columns

Figure 5.10.: Output from the isnull function applied to the iris flower data set.

It specifies whether each value in the dataframe is null (True) or not null (False). We would have to inspect all rows to verify the result. Note that the output of the isnull function is another pandas dataframe.

One way to verify that there are no null values in a column of the dataframe is to sum up the values of the corresponding column in the isnull dataframe. For example, the following code would check on the existence of null values in the SepalLengthCm column.

```
isnull_result = pd.isnull(df)
sepal_length_number_of_nulls = isnull_result['SepalLengthCm'].sum()
```

The value of sepal_length_number_of_nulls would come out to be 0. This works because the Boolean value False is interpreted as a 0 in the sum function. We can perform this check on all columns with the following

[21] 1 df.describe()

	Id	SepalLengthCm	SepalWidthCm	PetalLengthCm	PetalWidthCm
count	150.000000	150.000000	150.000000	150.000000	150.000000
mean	75.500000	5.843333	3.054000	3.758667	1.198667
std	43.445368	0.828066	0.433594	1.764420	0.763161
min	1.000000	4.300000	2.000000	1.000000	0.100000
25%	38.250000	5.100000	2.800000	1.600000	0.300000
50%	75.500000	5.800000	3.000000	4.350000	1.300000
75%	112.750000	6.400000	3.300000	5.100000	1.800000
max	150.000000	7.900000	4.400000	6.900000	2.500000

Figure 5.11.: Results from applying the describe function to the iris dataframe.

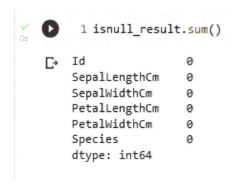

```
1 isnull_result.sum()
```

```
Id               0
SepalLengthCm    0
SepalWidthCm     0
PetalLengthCm    0
PetalWidthCm     0
Species          0
dtype: int64
```

We are fortunate, in this example, to be working with a clean data set.

Mine

To mine our data set requires performing some exploratory data analysis so that we can begin to discern any patterns in the data. Basic descriptive statistics on each column, representing a particular measured attribute, can be obtained with the pandas describe function. Figure 5.11 shows the result for the iris flower dataframe.

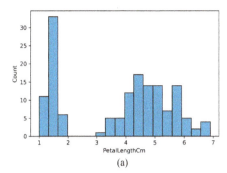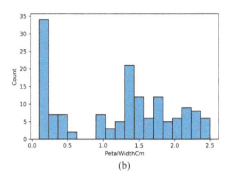

(a)   (b)

Figure 5.12.: Histograms for the petal measurements.

Looking at a histogram for each numerical observation can be an informative part of exploratory data analysis. The Seaborn module has a histogram function that will create professional-looking histograms. To generate a histogram of the petal length measurements, use the following

```
sns.histplot(df, x='PetalLengthCm' , bins=18)
```

Figure 5.12 shows the resulting histograms for both petal measurement. Note that there is an interesting double peak structure which may be related to the species.

When we have multivariate data to explore a pairwise or all-against-all scatterplot can be useful to identify possible correlations and clustering in the data. A quick way to produce such a plot when the data is in a pandas dataframe is to use the seaborn module pairplot function.

```
import seaborn as sns
g = sns.pairplot(df,hue="Species")
g.savefig('iris_pairplot.png' , dpi=300)
```

Figure 5.13 shows the resulting pairwise plot. The diagonal plots in the matrix are the frequency distributions for the variable indicated by the column or row label.

Looking at the pairwise plot suggests that the following plots might be useful in identifying species:

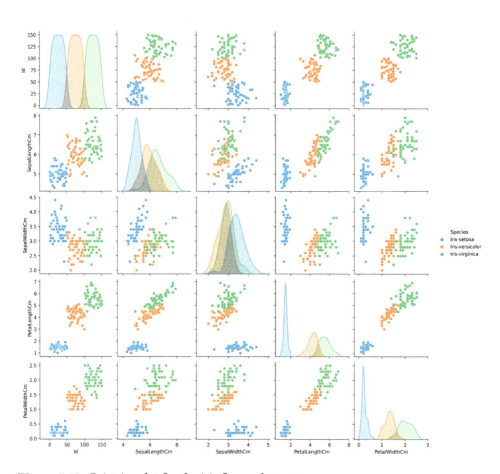

Figure 5.13.: Pairwise plot for the iris flower data set.

```
# Create a dataframe for each species
setosa = df.loc[df['Species']=='Iris-setosa']
versicolor = df.loc[df['Species']=='Iris-versicolor']
virginica = df.loc[df['Species']=='Iris-virginica']

# Create numpy arrays containing the numerical measurements
setosa_np = setosa.loc[:,['SepalLengthCm','SepalWidthCm','PetalLengthCm',
        'PetalWidthCm']].to_numpy()
versicolor_np = versicolor.loc[:,['SepalLengthCm','SepalWidthCm','
    PetalLengthCm',
        'PetalWidthCm']].to_numpy()
virginica_np = virginica.loc[:,['SepalLengthCm','SepalWidthCm',
            'PetalLengthCm', 'PetalWidthCm']].to_numpy()
```

Figure 5.14.: Code for creating dataframes for each species.

- petal length versus sepal length
- petal length versus sepal width
- petal length versus petal width

To extract the data required for producing these three plots using matplotlib we will create new pandas dataframes containing one species each and then extract the numerical columns and convert them to numpy arrays. Figure 5.14 shows the code that will accomplish this.

Represent

Based on the exploratory data analysis performed in the mining part of our work-flow we want to construct three scatter graphs that show petal lengths versus sepal lengths, sepal widths, and petal widths. We will use matplotlib so that we have greatest control over the final appearance of the plots. The code in 5.15 can be used for the petal length versus sepal length graph.

This code can be modified to produce the other two graphs by changing the column used for the y axis in the plot function. Figure 5.16 shows the three graphs created in this step of the workflow.

```
# Create Petal Length versus Sepal Length plot
plt.plot(setosa_np[:,2],setosa_np[:,0], linestyle='', marker='d',
         label='setosa')
plt.plot(versicolor_np[:,2],versicolor_np[:,0], linestyle='', marker='o',
         label='versicolor')
plt.plot(virginica_np[:,2],virginica_np[:,0], linestyle='', marker='^',
         label='virginica')
plt.legend()
plt.xlabel('Sepal Length [cm]')
plt.ylabel('Petal Length [cm]')
plt.title('Petal Length versus Sepal length')
plt.savefig('PetalLengthVsSepalLength.png', dpi=300)
```

Figure 5.15.: Code for creating the petal length versus sepal length plot.

Refine

We should refine our visualization to clarify the representation or by changing attributes that contribute to readability. One thing that might be changed in Figure 5.16 to aid comparing the graphs is to make the y-axis scale the same. We can set the y-axis scale with

```
plt.ylim((0,8))
```

The result of putting this statement in the code is Figure 5.17.

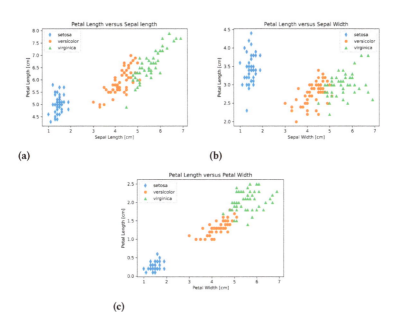

Figure 5.16.: Iris flower measurement comparisons by species.

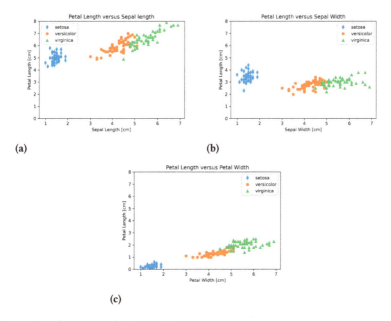

Figure 5.17.: Refinement of Figure 5.16

5.5. Computational Problem Solution: Comparing CO_2 Emissions by Country

As a final example of developing scientific visualizations, we return to the problem posed at the beginning of the chapter: looking at CO_2 emissions from fossil fuel use. We will focus on two questions.

1. How do nations of the world compare in their CO_2 emissions from fossil fuel use for a recent year?
2. How has the CO_2 emission from fossil fuel use changed over time for the highest emitters?

Our target audience will be policymakers in government.

Acquire - Obtain the data

We will use a subset of data from the U.S. Department of Energy (Boden et al., 2013). The subset is provided in the csv file fossil-fuel-co2-emissions-by-nation.csv. A description of each column in the file, including units used is in Table 5.2.

Parse or Understand the Data - Provide structure for the data's meaning, and order it into categories.

We will import the data as a Pandas dataframe and then look at information about the data set using the pandas info and describe functions. Screenshots of the Colab notebook that performs these operations are shown in Figure 5.18 and Figure 5.19

```
[1]  1 from google.colab import drive
     2 drive.mount('/content/drive')
     3 %cd//content/drive/MyDrive/Courses/CST210/Wentworth_Textbook/Drafts/Ch5/code

     Mounted at /content/drive
     /content/drive/MyDrive/Courses/CST210/Wentworth_Textbook/Drafts/Ch5/code


[2]  1 import numpy as np
     2 import matplotlib.pyplot as plt
     3 import pandas as pd
     4 import seaborn as sns


[3]  1 # Import data as pandas dataframe
     2 df = pd.read_csv('fossil-fuel-co2-emissions-by-nation.csv')


[4]  1 # Print out basic information about the dataframe
     2 df.info()

     <class 'pandas.core.frame.DataFrame'>
     RangeIndex: 17232 entries, 0 to 17231
     Data columns (total 10 columns):
      #   Column                    Non-Null Count  Dtype
     ---  ------                    --------------  -----
      0   Year                      17232 non-null  int64
      1   Country                   17232 non-null  object
      2   Total                     17232 non-null  int64
      3   Solid Fuel                17232 non-null  int64
      4   Liquid Fuel               17232 non-null  int64
      5   Gas Fuel                  17232 non-null  int64
      6   Cement                    17232 non-null  int64
      7   Gas Flaring               17232 non-null  int64
      8   Per Capita                17232 non-null  float64
      9   Bunker fuels (Not in Total) 17232 non-null int64
     dtypes: float64(1), int64(8), object(1)
     memory usage: 1.3+ MB
```

Figure 5.18.: Colab notebook section illustrating some parsing of the data.

```
[5]  1 df.describe()
```

	Year	Total	Solid Fuel	Liquid Fuel	Gas Fuel	Cement	Gas Flaring	Per Capita	Bunker fuels (Not in Total)
count	17232.000000	1.723200e+04	1.723200e+04	17232.000000	17232.000000	17232.000000	17232.000000	17232.000000	17232.000000
mean	1961.579561	2.268712e+04	1.107010e+04	7589.085829	3189.767700	638.453865	199.718489	0.907776	560.330606
std	44.251691	1.132419e+05	6.206518e+04	39057.928585	20714.456024	6631.010202	1087.880733	2.194268	2414.320487
min	1751.000000	-1.473000e+03	-1.030000e+02	-4663.000000	-40.000000	0.000000	0.000000	-0.680000	0.000000
25%	1944.000000	1.170000e+02	0.000000e+00	21.000000	0.000000	0.000000	0.000000	0.000000	0.000000
50%	1972.000000	9.645000e+02	5.400000e+01	263.000000	0.000000	7.000000	0.000000	0.130000	4.000000
75%	1995.000000	8.059250e+03	2.002500e+03	2165.750000	71.000000	162.000000	0.000000	1.010000	133.000000
max	2014.000000	2.806634e+06	2.045156e+06	680284.000000	390719.000000	338912.000000	20520.000000	45.950000	45630.000000

Figure 5.19.: Result of the pandas describe function.

One important observation from the describe results is that the final year for which we have data is 2014.

Next, we check for any null values in the dataset that would have to be removed using the pandas isnull function. The results are shown in Figure 5.20.

```
[6]    1 # Check for null values
       2 isnull_result = pd.isnull(df)
       3 isnull_result.sum()

Year                            0
Country                         0
Total                           0
Solid Fuel                      0
Liquid Fuel                     0
Gas Fuel                        0
Cement                          0
Gas Flaring                     0
Per Capita                      0
Bunker fuels (Not in Total)     0
dtype: int64
```

Figure 5.20.: Checking for null values in the data set.

Filter – Clean the data and remove all but the data of interest.

Again, we are fortunate in that the data set is relatively clean. There are no null entries that must be deleted to work with the plotting functions. It is likely that we will need to extract a subset of the data to focus on answering the questions that we posed. The data for 2014 can be extracted into its own dataframe with

```
# Extract rows for 2014
data_2014 = df.loc[(df['Year'] == 2014) & (df['Total']>5.e4)]
data_2014_sort = data_2014.sort_values('Total', ascending=False)
```

Mine – Use methods from statistics, data mining, or more fundamental scientific principles to discern patterns or place the data in a mathematical context.

Creating histograms for each measurement ends up not being very informative, except to show that most countries have very low emissions for most years. The high emitter countries dominate.

A basic pairwise or all-against-all scatterplot can be produced with the following code

```
g = sns.pairplot(df, corner=True)
g.savefig('CO2Emissions_pairplot.png', dpi=300)
```

The corner keyword argument is used to produce a plot containing the bottom left triangle of plots, since the top right triangle contains the same information with axes reversed. Figure 5.21 shows the result.

There are some interesting correlations suggested in Figure 5.20, but they are probably not relevant for answering our questions.

Represent - Choose a basic visual model, such as a bar graph, scatter graph, contour plot, or other visual construct.

A bar chart will be a good way of comparing country data for 2014, the most recent year for which we have data. We will use the matplotlib horizontal bar plot function, barh. We must extract the relevant columns from the dataframe and convert them to numpy arrays. The following code will generate an appropriate bar plot.

```
# Convert columns to numpy arrays
Country_data = data_2014_sort['Country'].to_numpy()
Total_data = data_2014_sort['Total'].to_numpy()
plt.barh(Country_data, Total_data)
plt.yticks(fontsize=6)
plt.tight_layout()
plt.savefig('CO2Emissions2014.png', dpi=300)
```

The result is Figure 5.22.

To address the question of how emissions have changed over time for the high emitter countries we will use a scatterplot that shows total emission in a year as a function of year. We will focus on the top five emitters: China, United States, India, Russian Federation, and Japan. Figure 5.23

Refine

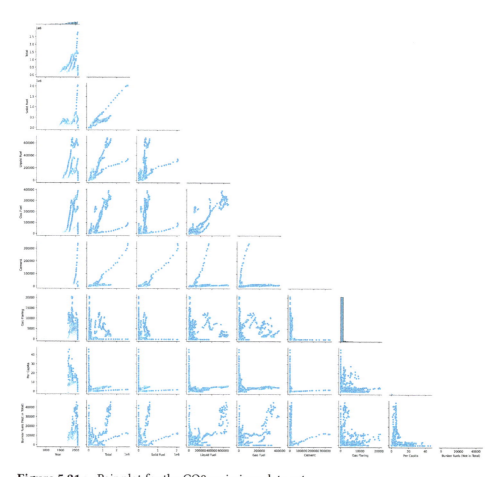

Figure 5.21.: . Pair plot for the CO2 emissions data set.

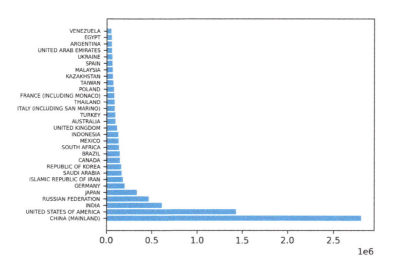

Figure 5.22.: . Comparison by country of CO$_2$ emission data for 2014.

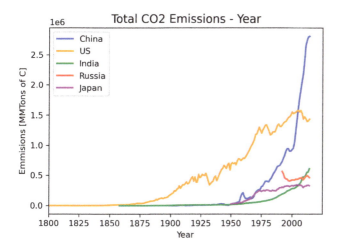

Figure 5.23.: CO$_2$ emissions from fossil fuels versus year.

The final version of the horizontal bar plot should have an x-axis label and a title. This is accomplished with the code in Figure 5.24. Note the additions in lines 31-33. The result is shown in Figure 5.25.

Figure 5.23 shows no measurable emissions for any of the displayed countries before 1850 on the chosen y-axis scale, therefore it would be better to define the x-axis limits to start with 1850. To show data for Russia over the same time period we would need to create a data series that combines the USSR with Russia. To avoid this complication, we will just delete Russia from the displayed countries. Since color will not always show for printouts of the graph, we will use grayscale colors, line width and line style changes to help distinguish the different countries. Figure 5.27 shows the code for the final form of the plot. Figure 5.26 shows the final plot.

```python
"""
Title: Fossil Fuel CO2 Emissions - 2014
Author: C.D. Wentworth
Version: 7.31.2022.2
Summary: This program will read in U.S. Dept. of Energy data for
         CO2 emissions by fossil fuel and produce a bar graph for
         the year 2014.

Revision History:
    7.31.2022.1: base
    7.31.2022.2: adjusts font size for the y-axis

"""
import numpy as np
import matplotlib.pyplot as plt
import pandas as pd
import seaborn as sns
# Import data as pandas dataframe
df = pd.read_csv('fossil-fuel-co2-emissions-by-nation.csv')
# Extract rows for 2014
data_2014 = df.loc[(df['Year'] == 2014) & (df['Total']>5.e4)]
data_2014_sort = data_2014.sort_values('Total', ascending=False)

# Convert columns to numpy arrays
Country_data = data_2014_sort['Country'].to_numpy()
Total_data = data_2014_sort['Total'].to_numpy()

# Create bar plot using matplotlib
plt.barh(Country_data, Total_data)
plt.xlabel('Emissions [MMTons of C]')
plt.title('Total CO2 Emissions for 2014', fontsize=14)
plt.yticks(fontsize=6)
plt.tight_layout()
plt.savefig('CO2Emissions2014V2.png', dpi=300)
plt.show()
```

Figure 5.24.: Final code for country comparison for 2014.

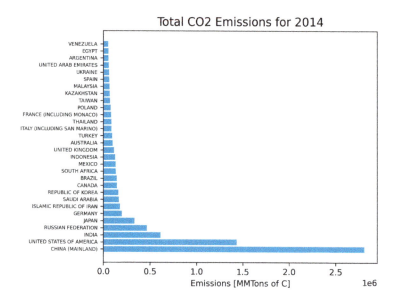

Figure 5.25.: Total CO2 emissions by country, final form.

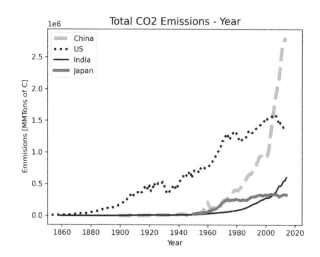

Figure 5.26.: CO2 emission versus year, final form.

```python
"""
Title: Fossil Fuel CO2 Emissions Versus Year
Author: C.D. Wentworth
Version: 7.31.2022.2
Summary: This program will read in  U.S. Dept. of Energy data for
         CO2 emissions by fossil fuel and produce a scatter graph for
         of emissions versus year.
Revision History:
    7.31.2022.1: base
    7.31.2022.2: change x-axis scale and linestyles

"""
import numpy as np
import matplotlib.pyplot as plt
import pandas as pd
import seaborn as sns

# Import data as pandas dataframe
df = pd.read_csv('fossil-fuel-co2-emissions-by-nation.csv')

# Extract rows of particular countries
China_data = df.loc[df['Country'] == 'CHINA (MAINLAND)']
US_data = df.loc[df['Country'] == 'UNITED STATES OF AMERICA']
India_data = df.loc[df['Country'] == 'INDIA']
Japan_data = df.loc[df['Country'] == 'JAPAN']
```

Figure 5.27a: Code for emissions versus year, final form.

```
26  # Convert columns to numpy arrays
27  China_data_year = China_data['Year'].to_numpy()
28  China_data_total = China_data['Total'].to_numpy()
29  US_data_year = US_data['Year'].to_numpy()
30  US_data_total = US_data['Total'].to_numpy()
31  India_data_year = India_data['Year'].to_numpy()
32  India_data_total = India_data['Total'].to_numpy()
33  Japan_data_year = Japan_data['Year'].to_numpy()
34  Japan_data_total = Japan_data['Total'].to_numpy()
35
36  # Create scatter plot - matplotlib
37  plt.plot(China_data_year, China_data_total, linewidth=5, color='silver',
38          linestyle='--', label='China')
39  plt.plot(US_data_year, US_data_total, linewidth=3, color='black',
40          linestyle=':', label='US')
41  plt.plot(India_data_year, India_data_total, linewidth=2, color='black',
42          label='India')
43  plt.plot(Japan_data_year, Japan_data_total, linewidth=4, color='gray',
44          label='Japan')
45  plt.xlim(xmin=1850)
46  plt.xlabel('Year')
47  plt.ylabel('Emmisions [MMTons of C]')
48  plt.title('Total CO2 Emissions - Year', fontsize=14)
49  plt.legend()
50  plt.savefig('CO2EmissionsVersusYearFinal.png', dpi=300)
51  plt.show()
```

Figure 5.27b: Continuation of code for emissions versus year, final form.

5.6. Exercises

1. True or False: Scientific visualization is the same thing as computer graphics.

2. Scientific visualization is the process of selecting and combining of data to help in discovering laws or in communicating results appropriately for a given audience.

3. Scientific visualizations can be classified using the following dimensions (or axes): (choose all that apply)

   a. content

b. dimensionality (2D/3D)

c. time

d. color

4. Classify the following visualization using the system in Figure 5.1.

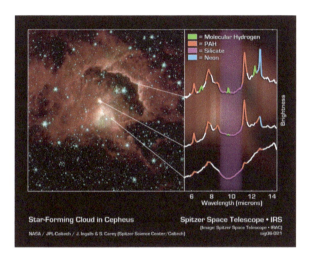

5. Classify the following visualization using the system in Figure 5.1.

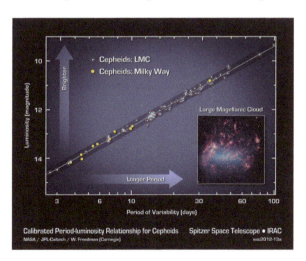

6. Classify the following visualization using the system in Figure 5.1. This visualization shows the hydrogen electron wave function for the n=4, l=3, m=0 state.

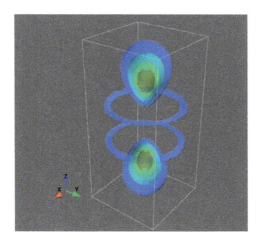

5.7. Program Modification Problems

1. The code shown below creates a histogram of the petal length measurements in the Iris Flower Data Set similar to Figure 5.13(a), except that bars have been colored according to species. You need to modify this code so that

■ It will create a histogram for the sepal length in the data set.

■ The x-axis label says "Sepal Length [cm]"

```
"""
Title: Iris Histogram
Author: C.D. Wentworth
Version: 7.31.2022.1
Summary: This program will read in the Iris data set and
         produce a histogram of one of the measurements.
Revision History:
    7.31.2022.1: base

"""

import numpy as np
import matplotlib.pyplot as plt
import pandas as pd
```

```
import seaborn as sns

# Import data as pandas dataframe
df = pd.read_csv('Iris.csv')

# Create a histogram - Seaborn version
sns.histplot(df, x='PetalLengthCm', bins=18, hue='Species')
plt.savefig('petalLengthHist.png', dpi=300)
plt.show()
```

2. The code below plots three exponential functions on the same graph. You need to modify the code so that it does the following:

■ plots the following three functions on the same graph:

$$f_1(x) = x \quad , \quad f_2(x) = x^2 \quad , \quad f_3(x) = x^3$$

■ plot the functions over the range $0.1 \leq x \leq 10$

■ Change the legend to indicate the power involved in each function:

'p = 1' , 'p = 2' , 'p = 3'

■ Make both the x-axis scale and the y-axis scale logarithmic.

```
"""
Title: Plotting Multiple Functions
Author: C.D. Wentworth
Version: 8.4.2022.1
Summary: This program will plot several functions on one graph
         using matplotlib.
Revision History:
         8.4.2022.1: base

"""
import matplotlib.pylab as plt
import numpy as np
x = np.linspace(0,5,40)
y1 = 0.10*np.exp(0.10*x)
y2 = 0.10*np.exp(0.20*x)
y3 = 0.10*np.exp(0.30*x)
plt.plot(x,y1,color='k',label='mu = 0.1',linestyle='solid', linewidth=3)
plt.plot(x,y2,color='b',label='mu = 0.2',linestyle='dashdot', linewidth=3)
plt.plot(x,y3,color='r',label='mu = 0.3',linestyle='dotted', linewidth=3)
```

```
plt.xlabel('x', fontsize=14)
plt.ylabel('y', fontsize=14)
plt.legend(loc='upper left')
plt.tight_layout()
plt.savefig('Ch5ProgModProb2.png', dpi=300)
plt.show()
```

3. The code shown below creates a scatterplot using the planets data. It uses the Seaborn module, as discussed in section 5.3.3. You need to modify the code to do the following

■ Use data for bacterial growth in the file BacterialGrowthData.txt. This data gives the measured bacterial cell density in the growth bottle filled with either CHSA or TSB media (liquid food) as a function of time. To read this data into a Pandas dataframe you will need to tell read_csv that the file is tab-delimited instead of comma-delimited. This is done with sep keyword argument:

sep='\t'

You will also need to inspect the data file to determine which row is the header that contains the column headings. Remember that Python uses zero-based indexing.

■ Create a scatterplot of the CHSA column versus time.

■ Use the whitegrid Seaborn style.

■ Set the x-axis label to 't [min]'

■ Set the y-axis label to ' N [rel]'. The N stands for the density of bacteria in the sample. The units [rel] indicate relative units.

■ Add a title: 'Bacterial Growth versus Time: CHSA Media' ; set the font size to 14.

■ Save the graph as a png graphics file.

```
"""
Title: Scatter Plot - Seaborn Version
Author: C.D. Wentworth
Version: 8.4.2022.1
Summary: This program will plot several functions on one graph
         using matplotlib.
```

```
Revision History:
        8.4.2022.1: base

"""
import matplotlib.pyplot as plt
import numpy as np
import pandas as pd
import seaborn as sns
sns.set_style('darkgrid')

# Read in exoplanet data as a pandas dataframe
planets = pd.read_csv('PSCompPars_subset.csv', sep=',', header=16)

# Create a scatterplot
sns.lmplot(x='pl_orbsmax', y='pl_orbper', data=planets,  fit_reg=False)
plt.xlim(0,1.2)
plt.xlabel('Orbital Semi-major Axis [au]')
plt.ylabel('Orbital Period [days]')
plt.savefig('pl_orbsmax-pl_orbper.png', dpi=300)
plt.show()
```

5.8. Program Development Problems

1. Develop visualizations that will help explore how per capita fossil fuel CO2 emissions have developed over time and how they compare by country in recent years. You can use the data in the csv file

fossil-fuel-co2-emissions-by-nation.csv

Go through the visualization development workflow discussed in section 5.4. In addition to the code file or Jupyter notebook that you create, you need to write a brief summary of how you executed each workflow step (excluding the addition of interaction).

5.9. References

Atkinson, J. (2017, April 10). *What is Earth's Energy Budget? Five Questions with a Guy Who Knows* [Text]. NASA. http://www.nasa.gov/feature/langley/what-is-earth-s-energy-budget-five-questions-with-a-guy-who-knows

Ausoni, C. O., Frey, P., & Tierny, J. (2014). *Scientific Visualization at the interfaces.* https://www.ljll.math.upmc.fr/frey/visu.html

Boden, T. A., Andres, R. J., & Marland, G. (2013). *Global, Regional, and National Fossil-Fuel CO2 Emissions (1751—2010) (V. 2013).* Environmental System Science Data Infrastructure for a Virtual Ecosystem (ESS-DIVE) (United States); Carbon Dioxide Information Analysis Center (CDIAC), Oak Ridge National Laboratory (ORNL), Oak Ridge, TN (United States). https://doi.org/10.3334/CDIAC/00001_V2013

Fisher, R. A. (1936). The Use of Multiple Measurements in Taxonomic Problems. *Annals of Eugenics, 7*(2), 179–188. https://doi.org/10.1111/j.1469-1809.1936.tb02137.x

Fry, Ben. (2008). *Visualizing data.* O'Reilly Media, Inc.; WorldCat.org. https://www.loc.gov/catdir/toc/fy0804/2008297507.html

Iris Species. (n.d.). Retrieved July 31, 2022, from https://www.kaggle.com/datasets/uciml/iris

List of named colors—Matplotlib 3.5.2 documentation. (2022). https://matplotlib.org/stable/gallery/color/named_colors.html

NASA Exoplanet Archive. (n.d.). Retrieved August 2, 2022, from https://exoplanetarchive.ipac.caltech.edu/index.html

pandas.read_csv—Pandas 1.4.3 documentation. (n.d.). Retrieved August 1, 2022, from https://pandas.pydata.org/docs/reference/api/pandas.read_csv.html

Power Project Team. (n.d.). *NASA POWER | Prediction Of Worldwide Energy Resources*. Retrieved July 30, 2022, from https://power.larc.nasa.gov/

Rathod, V. (2020). *Iris Flower CaseStudy*. RPubs. https://rpubs.com/vidhividhi/irisdataeda

Shinker, J. J. (2016). *Global Climate Animations*. Gobal Climate Animations. http://climvis.org/content/global.htm

Shneiderman, B. (2003). The Eyes Have It: A Task by Data Type Taxonomy for Information Visualizations. In B. B. Bederson & B. Shneiderman (Eds.), *The Craft of Information Visualization* (pp. 364–371). Morgan Kaufmann. https://doi.org/10.1016/B978-155860915-0/50046-9

The pandas development team. (2022). *pandas—Python Data Analysis Library*. https://pandas.pydata.org/

Tory, M., & Moller, T. (2004). Rethinking Visualization: A High-Level Taxonomy. *IEEE Symposium on Information Visualization*, 151–158. https://doi.org/10.1109/INFVIS.2004.59

6. Functions

Computer applications used in science and engineering can involve thousands or millions of lines of code. Faced with the complexity of such programs, programmers can use the computational science pillars of abstraction and decomposition to create a high-level conceptualization of the program and then break the coding into more manageable pieces. One method of applying the ideas of abstraction and decomposition is to use named blocks of code or program routines. In Python, program routines are called functions.

Functions will aid in managing code complexity and achieving code modularity: creating code that can be used to solve multiple computational problems. In this chapter, we will learn how to create user-defined functions and become familiar with underlying computer science concepts that will help us use functions correctly in a program.

Motivating Problem: Temperature Conversion –Revised

In Chapter 3, we developed a program to perform user-requested temperature conversions, such as going from Fahrenheit to Celsius. While the code structure is evident due to added comments, we can improve the clarity by defining functions to perform parts of the program, such as requesting a temperature from the user and performing the temperature conversion. Having functions for these tasks will allow us to reuse some of the code in other programs. Therefore, the problem we want to solve is to rewrite the code from Chapter 3 so that the following tasks are performed in their own separate functions:

- request a temperature to convert from the user
- request the final temperature scale to be used
- perform the actual temperature conversion calculation

6.1. Program Routines

A computer program can contain a block of code that gets repeated. We have seen that iterative control structures do this. If the block contains many lines of code, then the program's design can be clarified by bundling the block into a separate part that a single line of code can substitute. A block of code that a name can reference is called a **program routine**. A routine can be called up and executed as often as required by just using the name. Figure 6.1 shows the concept of reusing a named block of code. Figure 6.2 shows how a code block in a program can be replaced by a single code line using a program routine. Program routines are called functions in Python.

Main Program

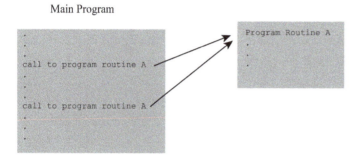

Figure 6.1.: Illustrating the use of a Program Routine.

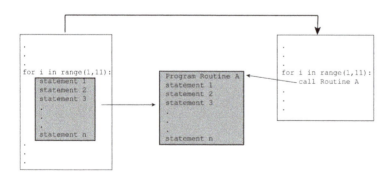

Figure 6.2.: Replacing a code block with a routine to clarify program structure.

6.2. Python Functions as a Program Routine

The basic structure of a Python function definition is

```
def function_name(p1, p2, p3):
    statement 1
    statement 2
    .
    .
    .
```

The function definition begins with the header statement def, and the function name can be any legal Python variable name. The variable names in the parentheses are called parameters, which represent values that will be passed to the function when it is called. All statements inside the function definition comprise a suite (or block) of code and must all be indented by the same amount.

The parameter names, p1, p2, and p3, above are simply placeholders. They will be assigned actual values when the function is called from the main program. Consider the following example.

```
var1 = 10
var2 = 20
var3 = 30
function_name(var1, var2, var3)
```

When the function is called, the parameters p1, p2, and p3 get replaced by the values in the variables var1, var2, and var3. Values that get passed to a function when it is called are arguments. The placeholder names appear in the function definition, and those placeholders get replaced by corresponding argument values when the function is called.

Python functions must be defined before they are called in a statement. Function definitions typically appear at the top of a program.

6.2.1. Value-returning Functions

In mathematics, the function notation, $f(x)$ indicates a value that depends on the value of x for its definition. If you see $f(x)$ in a mathematical statement, you can imagine it being replaced by the appropriate value. Similarly, value-returning

```
✓ [1]    1 def greeting(name):
 0s       2     print('Hi', name)
          3     print("It's nice to see you.")
          4     print('I hope you will visit again.')


✓ [2]    1 greeting('Tom')
 0s

         Hi Tom
         It's nice to see you.
         I hope you will visit again.
```

Figure 6.3.: Non-value returning function.

functions get replaced by a value generated by the function in a statement containing the function. The returned value is indicated in the function definition through the use of the keyword return. The following code shows an example of a value-returning function.

```
def factorial(n):
    f = 1
    for i in range(1,n+1):
        f = f*i
    return f

# Main Program
print(factorial(4))
```

The output of the print function is 24 in this example.

6.2.2. Non-value-returning Functions

As the name suggests, a non-value-returning function does not return a value when it is called in a statement; instead, it is called because of **side effects** that occur when the function is executed. For example, a non-value returning function might cause specific messages to be printed out. Figure 6.3 shows an example. Note that the function greeting(name) has no return statement, but it still has an effect.

Strictly speaking, the non-value-returning function will return a value: None, but we will treat such functions as truly non-value-returning.

6.3. Function Parameters and Arguments

6.3.1. Parameter Passing

The variable names used in the parameter list of a function definition are placeholders. They get replaced by actual argument values when the function is called in a statement. Consider the following simple function.

```
def f(n1, n2):
    t = 3.1415*(n1 + 2.0*n2)
    return t
```

The main program contains the following statements

```
v1 = 3.0
v2 = 4.0
print(f(v1, v2))
```

When the function f is called in the print statement, the parameters n1 and n2 in the function definition get replaced by the arguments v1 and v2. The call to f in the print statement results in the display of the number 34.56. The parameters are replaced by the arguments based on the order. If later in the main program, the following code is executed

```
print(f(v2,v1)
```

the number 31.42 will be displayed.

6.3.2. Keyword Arguments

In the above example, values were assigned to function parameters using the position of the arguments in a function call. The technical term for such arguments is **positional arguments**. Values can also be assigned to parameters using parameter names. This approach is called **keyword arguments**. For the function *f* defined above, the following call to the function would result in 34.56 being displayed.

```
v1 = 3.0
```

```
v2 = 4.0
print(f(n2=v2, n1=v1))
```

If you use both positional and keyword arguments to assign values to parameters in a function call, the positional arguments must all be assigned first.

6.3.3. Default Arguments

Default values can be assigned to a function parameter when the function is defined. The following code gives an example.

```
def f(n1=3, n2=4):
    t = 3.1415*(n1 + 2.0*n2)
    return t
print(f())
```

The code will display 34.56 even though no arguments were provided to the function.

6.3.4. Variable Scope

The **scope** of a variable is the block of code in which the variable is defined, can be used, and can be modified. In a Python program that contains function definitions, variables can have either local or global scope. A local variable, or a variable with **local scope**, is defined and used within a function. A global variable, or a variable with **global scope**, is defined outside of a function. To explore the difference, consider the following short program.

```
def f1(n):
    t1 = 2*n
    t2 = t1 + 3
    return t2
def f2(n):
    t3 = 2*n + global_v
    return t3
global_v = 5
print(f1(global_v))
print(f2(global_v))
```

The first print statement displays the number 13, and the second print statement displays 15. Note that the variable global_v was not passed to function f2, but since it was defined outside of a function, it will have the value five everywhere in the program. global_v is a global variable or has global scope.

Now, after executing the above code, suppose we execute

```
print(t1)
```

The result will be an execution error because the variable t1 was defined in function f1, not outside it, so it is available only within f1. t1 is a local variable or has local scope.

6.4. Computational Problem Solving: Temperature Conversion – Function Version

The problem described at the beginning of the chapter was to revise the temperature conversion program from Chapter 3 so that the following pieces were performed in functions rather than in the main program.

- request a temperature to convert from the user
- request the final temperature scale to be used
- perform the actual temperature conversion calculation

6.4.1. Analysis

The critical analysis for this problem was performed in Chapter 3, where we collected all required conversion equations, Equations 3.1a-3.1f. We can use the test data in Table 3.5 to determine if our revised program is working correctly.

6.4.2. Design

We provide a revised pseudocode version of the temperature conversion program in Figure 6.4.

Data structures used in the revised program are listed in Table 6.1.

The revised pseudocode for the program contains three functions. We must create

```
PROGRAM temperature_conversion
    Print a program greeting
    submitted_temperature, submitted_scale = request_temperature()
    converted_scale = request_converted_scale(submitted_scale)
    converted_temperature =
            calculate_converted_temp(submitted_temperature,
            submitted_scale, converted_scale)
    Print out converted_temperature
ENDPROGRAM
```

Figure 6.4.: Pseudocode for the revised temperature conversion program.

Table 6.1.: Data structures (variables) required by the revised program.

Data Structure	Type	Description
submitted_temperature	float variable	temperature submitted by the user
submitted_scale	string variable	scale for submitted_temperature
converted_temperature	float variable	the converted temperature
converted_scale	string variable	scale for converted_temperature

pseudocode versions for each of those functions. These are contained in Figures 6.5, 6.6, and 6.7.

6.4.3. Implementation

The Python implementation of the three functions described above is straightforward and can be seen in Figures 6.8, 6.9, and 6.10. With those function definitions, the main program becomes short and very clear in its structure, as shown in Figure 6.11.

6.4.4. Testing

Test data listed in Table 3.5 will be used to test the revised program. The complete code for the revised program is contained in the file Ch6TempConvProg.py. The test results for the program are listed in Table 6.2. The results provide evidence that the program works according to the requirements.

```
FUNCTION request_temperature
    INPUT: None
    submission_is_incorrect = True
    scale_request = "'C' for Celsius, 'F' for Fahrenheit,
                    'K' for Kelvin "
    WHILE submission_is_incorrect DO
        Print request for temperature
        Get temperature and convert to float (defines
                                submitted_temperature)
        print request for scale
        get scale (defines submitted_scale)
        IF ((submitted_scale == 'C') and
            (submitted_temperature >= -273.15))
            submission_is_incorrect = False
        ELSE IF ((submitted_scale == 'F') and
                (submitted_temperature >= -459.67))
            submission_is_incorrect = False
        ELSE IF ((submitted_scale == 'K') and
                (submitted_temperature >= 0)):
            submission_is_incorrect = False
        ELSE
            print('Incorrect submitted temperature. Try again.')
        ENDIF
    ENDWHILE
    OUTPUT submitted_temperature, submitted_scale
ENDFUNCTION
```

Figure 6.5.: Pseudocode version of request_temperature function.

```
FUNCTION request_converted_scale
    INPUT: submitted_scale
    submission_is_incorrect = True
    WHILE submission_is_incorrect DO
        Print the request for the temperature scale
        Get scale from user (defines converted_scale)
        IF ((submitted_scale == 'C') and
                (converted_scale == 'F' or converted_scale == 'K'))
            submission_is_incorrect = False
        ELIF ((submitted_scale == 'F') and
                (converted_scale == 'C' or converted_scale == 'K'))
            submission_is_incorrect = False
        ELIF ((submitted_scale == 'K') and
                (converted_scale == 'C' or converted_scale == 'F'))
            submission_is_incorrect = False
        ELSE
            print('There is a problem with your submission.')
        ENDIF
    ENDWHILE
    OUTPUT converted_scale
ENDFUNCTION
```

Figure 6.6.: Pseudocode version of request_converted_scale function.

Table 6.2.: Test results for the Chapter 6 Temperature Conversion Program.

Submitted Temperature		Converted Temperature		
T	Scale	T	Scale	Program Output
-300	C	Error		error detected
-150	C	-238	F	-238
-150	C	123.15	K	123.15
0	C	32	F	32
0	C	273.15	K	273.15
-500	F	Error		error detected
-400	F	-240	C	-240
-400	F	33.15	K	33.15
0	F	-17.78	C	-17.78
0	F	255.37	K	255.37
-100	K	Error		error detected
0	K	-273.15	C	-273.15
0	K	-459.67	F	-459.67
150	K	-123.15	C	-123.15
150	K	-189.67	F	-189.67

```
FUNCTION calculate_converted_temp
    INPUT: submitted_temperature, submitted_scale, converted_scale
    IF submitted_scale == 'C':
        IF converted_scale == 'F'
            # convert Celsius to Fahrenheit
            converted_temperature = submitted_temperature*9.0/5.0 +
                                    32.0
        ELSE
            # convert Celsius to Kelvin
            converted_temperature = submitted_temperature + 273.15
        ENDIF
    ELIF submitted_scale == 'F':
        IF converted_scale == 'C'
            # convert Fahrenheit to Celsius
            converted_temperature = (submitted_temperature -
                                    32.0)*5./9.
        ELSE
            # convert Fahrenheit to Kelvin
            converted_temperature = (submitted_temperature -
                                    32.0)*5.0/9.0 + 273.15
        ENDIF
    ELSE
        IF converted_scale == 'C'
            # convert Kelvin to Celsius
            converted_temperature = submitted_temperature - 273.15
        ELSE
            # convert Kelvin to Fahrenheit
            converted_temperature = (submitted_temperature -
                                    273.15)*9.0/5.0 + 32.0
        ENDIF
    ENDIF
    OUTPUT converted_temperature
ENDFUNCTION
```

Figure 6.7.: Pseudocode version of calculate_converted_temp function.

```python
def request_temperature():
    scale_request = "'C' for Celsius, 'F' for Fahrenheit, 'K' for Kelvin "
    submission_is_incorrect = True
    while submission_is_incorrect:
        submitted_temperature = float(input('Submit a temperature: '))
        print('Specify the scale of your submitted temperature: ')
        submitted_scale = input(scale_request)
        if ((submitted_scale == 'C') and (submitted_temperature >= -273.15)
):
            submission_is_incorrect = False
        elif ((submitted_scale == 'F') and (submitted_temperature >=
-459.67)):
            submission_is_incorrect = False
        elif ((submitted_scale == 'K') and (submitted_temperature >= 0)):
            submission_is_incorrect = False
        else:
            print('Incorrect submitted temperature. Try again.')
    return submitted_temperature, submitted_scale
```

Figure 6.8.: Python implementation of request_temperature function.

```python
def request_scale(submitted_scale):
    print('What scale should be used for the converted temperature?')
    scale_request = "'C' for Celsius, 'F' for Fahrenheit, 'K' for Kelvin: "
    converted_scale = input(scale_request)
    submission_is_incorrect = True
    while submission_is_incorrect:
        if ((submitted_scale == 'C') and
            (converted_scale == 'F' or converted_scale == 'K')):
            submission_is_incorrect = False
        elif ((submitted_scale == 'F') and
            (converted_scale == 'C' or converted_scale == 'K')):
            submission_is_incorrect = False
        elif ((submitted_scale == 'K') and
            (converted_scale == 'C' or converted_scale == 'F')):
            submission_is_incorrect = False
        else:
            print('There is a problem with your submission.')
            converted_scale = input(scale_request)
    return converted_scale
```

Figure 6.9.: Python implementation of request_converted_scale function.

```python
def calculate_converted_temp(submitted_temperature, submitted_scale,
                             converted_scale):
    if submitted_scale == 'C':
        if converted_scale == 'F':
            # convert Celsius to Fahrenheit
            converted_temperature = submitted_temperature*9.0/5.0 + 32.0
        else:
            # convert Celsius to Kelvin
            converted_temperature = submitted_temperature + 273.15
    elif submitted_scale == 'F':
        if converted_scale == 'C':
            # convert Fahrenheit to Celsius
            converted_temperature = (submitted_temperature - 32.0)*5./9.
        else:
            # convert Fahrenheit to Kelvin
            converted_temperature = (submitted_temperature - 32.0)*5.0/9.0
+ 273.15
    else:
        if converted_scale == 'C':
            # convert Kelvin to Celsius
            converted_temperature = submitted_temperature - 273.15
        else:
            # convert Kelvin to Fahrenheit
            converted_temperature = (submitted_temperature - 273.15)
*9.0/5.0 + 32.0
    return converted_temperature
```

Figure 6.10.: Python implementation of calculate_converted_temp function.

```python
# Main Program

# Display program greeting
print('Welcome to the Temperature Scale Conversion Program!')
print('This program will request that the user submit a temperture.')
print('Next, it requests the converted scale.')
print('Finally, it prints out the converted temperature.')
submitted_temperature, submitted_scale = request_temperature()
converted_scale = request_scale(submitted_scale)
converted_temperature = calculate_converted_temp(submitted_temperature,
                                                 submitted_scale,
    converted_scale)
# print out the result
s1 = format(submitted_temperature, '.2f')
s2 = format(converted_temperature, '.2f')
print(s1, submitted_scale, ' is ', s2, converted_scale)
```

Figure 6.11.: Python implementation of the main program.

6.5. Exercises

1. T/F: A user-defined function can be called as many times as is necessary in a program.

2. When a function is defined, the variable names appearing in the parentheses are called _____.

3. When a function is called in the main part of a program, the values passed to it are called _____.

4. What kind of function must contain a return statement?

  a. any function

  b. value-returning function

  c. non-value-returning function

5. A non-value-returning function can perform useful work through _____-____.

6. T/F: A value-returning function can only return one value at a time.

7. When a function call contains no keyword arguments, then the correspondence between arguments and function parameters is determined by the _____ of the arguments.

8. A program routine is

  a. the identifier in a function header

  b. a named group of instructions in a program

  c. a built-in function of a programming language

d. the arguments passed from a function call

9. A formal parameter in a function is

a. a placeholder name used in the function header, inside parentheses

b. a value passed to the function in a function call

c. the keyword def in a function definition

d. a variable name that appears in the calling program for a function

10. Consider the following code

```python
def a_function(p1, p2):
    t = p1*p2
    return t

# Main Program
num1 = 10.0
print(a_function(num1, 20.0)
```

What is the scope of variable t in a_function?

6.6. Program Modification Problems

1. One of the main techniques used in developing a computational solution to a problem is abstraction. The primary technique for implementing abstraction in code is to define functions that solve a small part of the problem. Defining functions is especially useful when a certain process must be repeated many times across more than one program. The function definition can be reused in multiple programs.

This exercise will give you some practice in defining your own Python function.

You need to define a Python function that accepts an integer N as an argument and then adds all the integers from 1 to N and returns the value. The main program should ask the user for an integer, and then it should print out the result of the function. The code shown below is a starting point for your solution. Make sure to

■ Change the welcome message so that it is appropriate to the problem.

■ Define the function that calculates that does the calculation.

■ Prints out the user-defined integer and the sum with an appropriate message. This should be done in the Main Program.

```
"""
Program Name: Chapter 6 Prog Mod Prob 1
Author: C.D. Wentworth
version: 8.10.2022.1
Summary: This program will request that the user enter
        a positive integer and prints it out.
"""

# Function definition

# Main Program
# Display program welcome
print('This program will request a number from the user.')

# Get an integer
user_number = int(input('Enter a positive integer: '))

while (user_number-int(user_number)) != 0 or (user_number <= 0):
    user_number = int(input("Please enter a positive integer: "))

print('You entered N =', user_number)
```

2. Modify the Chapter 6 Temperature Conversion Program discussed in section ?? to include the Rankine scale in addition to Celsius, Fahrenheit, and Kelvin scales. Start with the code in the file

Ch6TempConversionProgram.py.

6.7. Program Development Problems

1. Write a program to produce a table of equivalent temperatures using the Kelvin, Celsius, Fahrenheit, and Rankine scales. Produce the columns in the order just described. Start with 0 [K] and go up to 375 [K] with a 5-degree increment. The numbers should be formatted to have two decimal places. Label each of the columns appropriately. The table should be written out to a tab-delimited text file.

7. Dynamical Systems Modeling I

A dynamical systems model will allow a researcher to make quantitative predictions about a system, particularly with respect to the time development of important system properties. Such models are based on fundamental principles, not just empirical data, so they contribute to our basic understanding of the system being investigated. With a dynamical systems model we can make more reliable predictions about the course of a disease in a population, the metabolism of a drug in the human body, how the concentration of CO_2 in the atmosphere changes over time, the motion of a space vehicle on its way to Mars, and even the motion of planets in our solar system over thousands of years.

This chapter will introduce you to some basic concepts of calculus that will make dynamical systems models easier to understand and some computational methods of solving the models to obtain predictions without having to know a lot of sophisticated mathematics.

Motivating Problem: Mathematical Model for Bacterial Growth

Pseudomonas aeruginosa is a common species of bacteria found in soil, water reservoirs polluted by animals and humans, the human gastrointestinal tract, and on human skin (Diggle & Whiteley, 2020). It is involved in opportunistic infections of people, often in hospital settings, which makes it an important microbe to study and understand. If a container of liquid growth medium is inoculated with a small sample of the microbe, then growth data such as that shown in Figure 7.1 will be obtained.

Our understanding of any microbial organism, such as *Pseudomonas aeruginosa*, will be aided by having an accurate mathematical model for the growth. The concepts developed in this chapter will help us to develop such a model and to simulate the model through numerical calculations done by computer.

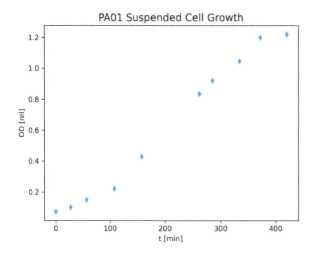

Figure 7.1.: Growth data for PA01 in 0.25% glucose medium.

7.1. Calculus Concepts

Your understanding of mathematical models will be aided by knowing just a couple of ideas from calculus, namely, the basic idea of a derivative of a function and the integral of a function. The goal here is not to become proficient at doing all the operations of taking derivatives or calculating integrals but to just have a basic conceptual understanding of what these mathematical things represent.

7.1.1. Derivatives

We will start with the derivative of a function. Consider the simple physics experiment of dropping a ball from some height and recording its position as a function of time. Table 7.1 gives some actual data taken from a video of such a ball drop. Figure 7.2 shows a graph of the data.

Table 7.1.: Height as a function of time for a dropped ball.

t [s]	y [m]
0.06673	2.693
0.1001	2.682
0.1335	2.67
0.1668	2.611
0.2002	2.576
0.2336	2.505
0.2669	2.446
0.3003	2.341
0.3337	2.235
0.367	2.141
0.4004	1.999
0.4338	1.882
0.4671	1.729
0.5005	1.588
0.5339	1.411
0.5672	1.247
0.6006	1.047
0.634	0.8586
0.6673	0.6586
0.7007	0.4705
0.7341	0.2588

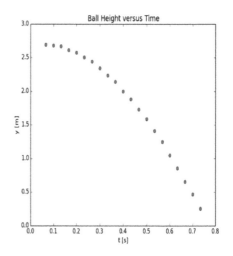

Figure 7.2.: Ball height as a function of time. Data is from Table 7.1

You can probably tell from the graph that the ball is speeding up over time, as it drops. The data points are at equal time intervals and the change in y clearly gets bigger as time increases. We will describe the y-coordinate by the function $y(t)$. We define the average rate of change of $y(t)$ by

$$\overline{v}_y = \frac{y(t_2) - y(t_1)}{t_2 - t_1} = \frac{\Delta y}{\Delta t} \tag{7.1}$$

We choose to symbolize this number by $\overline{v}_y$ because in this case it turns out to be the average velocity of the ball, hence the v. We can easily estimate the average rate of change of $y(t)$ by using the data in the table. For the average rate of change between 0.2002 and 0.2669 [s] we have

$$\bar{v}_y = \frac{y(0.2669) - y(0.2002)}{0.2669 - 0.2002} = \frac{(2.446 - 2.576)\,[m]}{0.0667\,[s]} = -1.95\,[m/s] \qquad (7.2)$$

The main conceptual idea of a derivative of a function such as $y(t)$ is to calculate the average rate of change of $y(t)$ as the time interval gets infinitesimally small. The mathematical process of doing this is taking a limit, although you do not need to worry about the details of actually doing this. We call this value the instantaneous rate of change of $y(t)$ at a particular value of t. The symbol for the instantaneous rate of change of with $y(t)$ respect to t at $t = a$ is

$$\left.\frac{dy(t)}{dt}\right|_{t=a} \qquad (7.3)$$

and this quantity is called the derivative of $y(t)$ at $t=a$. Formally, we define

$$\left.\frac{dy(t)}{dt}\right|_{t=a} = \lim_{\Delta t \to 0} \frac{y(a + \Delta t) - y(a)}{\Delta t} \qquad (7.4)$$

If you take a calculus course, you will learn a variety of techniques for performing the limit procedure. For our purposes, we can interpret the derivative geometrically. The derivative of $y(t)$ at $t = a$ is the slope of the tangent line to $y(t)$ drawn at $t = a$. For example, Figure 7.3 shows a model calculation for the ball height $y(t)$ with an estimate of the derivative of $y(t)$ at $t = 0.20$ [s]. A straight line is drawn tangent to the curve at $t = 0.20$ [s], and the slope of the straight line is estimated from the rise and run. We get

$$\left.\frac{dy(t)}{dt}\right|_{t=0.20} = \text{slope of tangent line} = \frac{-1.69\,[m]}{0.80\,[s]} = -2.1\,[m/s] \qquad (7.5)$$

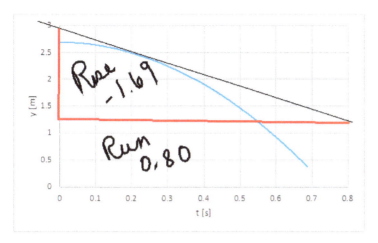

Figure 7.3.: Illustration of finding the derivative of a function at t = 0.2 [s] by estimating the slope of the tangent line.

7.1.2. Integrals

The second calculus concept we need is the integral of a function. The geometrical meaning of the definite integral of a function $v(t)$ between t_1 and t_2 is the area under the curve between $t=t_1$ and $t=t_2$

In Figure 7.4, the shaded area represents the integral of $v(t)$ between t_1 and t_2.

Mathematicians use a special symbol to represent the integral of a function, as shown in the following equation.

$$\int_{t_1}^{t_2} v\,(t)\,dt = \text{area under v-t graph from } t_1 \text{ to } t_2 \qquad (7.6)$$

As an example, consider Figure 7.5. Let's calculate the integral of $v(t)$ from t = 0 to t = 2 [s].

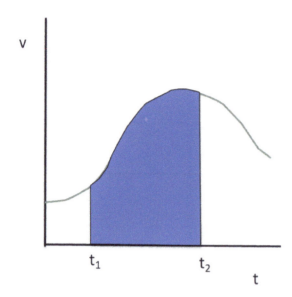

Figure 7.4.: The shaded area represents the integral of v(t) between t1 and t2 .

$$\int_0^2 v(t)\, dt = \text{area under v-t graph from } 0[s] \text{ to } 2[s] = \frac{1}{2}\text{base} \times \text{height} \quad (7.7)$$

$$= \frac{1}{2} \times 2 \times 20$$

$$= 20 \left[m^2 \right]$$

The integral of a function is essentially the opposite of the derivative. In fact, if you integrate the derivative of a function you end up with the function itself. We can express this as

$$\int_{t_1}^{t_2} \frac{dy}{dt}\, dt = y(t_2) - y(t_1) \quad (7.8)$$

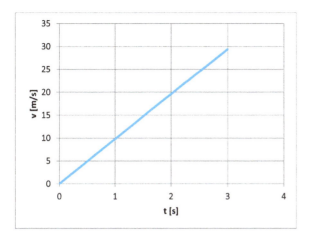

Figure 7.5.: Velocity as a function of time for a particle with constant acceleration.

7.1.3. Numerical Calculation of Derivatives and Integrals

Both the derivative of a function and the integral of a function can be calculated numerically with standard Python library functions. The derivative function is in the scipy sublibrary named misc (The SciPy community, 2022b). A template for using the derivative function is shown below.

```
import scipy.misc as spm
def f(x):
    return x**3
dfx = spm.derivative(f,2,dx=1e-6,n=1)
```

The scipy library scipy.integrate has several functions for performing numerical integration. They differ by the algorithm used. The best first choice is the quad function (The SciPy community, 2022a). Here is the basic syntax for using the quad function.

```
scipy.integrate.quad(f,a,b,args=())
```

f : a Python function

a: float that is the lower limit of integral

b: float that is the upper limit of integral

args: a tuple that contains any parameters required by f

The quad function returns a tuple: (value of integral, error estimate). A template for using the quad function to perform integration is shown below.

```
import scipy.integrate as si
def f(t):
    return (30. - 1.8*t - 0.040*t**2)
intFunc = si.quad(f,0,5)
integral_value = intFunc[0]
```

7.2. Definition of A Dynamical System

A dynamical system is one that changes with time. Mathematically, it is a set of variables that describe the system of interest, and these variables will all depend on time. The state of the system is described by the value of the variables at a particular time. The present state of the system will depend on the system state in the past. The mathematical model defining the dynamical system is typically described by the state variable definitions and equations describing the rate of change for these variables.

Examples:

- Population of an organism
- electrical behavior of a network of neurons
- Planetary positions in a stellar system
- the motion of molecules in a fluid
- drug concentration in a body part
- many, many more biological, physical, and social systems

In the simplest case, the state of the system will be described by the value of one variable, which will be a function of time. Let us represent the state by the variable $y(t)$, which is shown explicitly to be a function of time. y might be concentration of bacteria in a container, for example. For this one-variable system, the model is specified by the time-rate-of-change of y, which is just the derivative of y with respect to t.

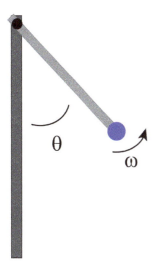

Figure 7.6.: Pendulum geometry.

$$\frac{dy}{dt} = f(t, y, \mathbf{p}) \tag{7.9}$$

The function f is assumed to be known. We see that it can depend on the current time t, the current value of the state variable y, and on model parameters contained in the vector $\mathbf{p}$. If the value of y is known at one time, often $t = 0$, then Equation will determine the behavior of y for all subsequent t. The general term for this kind of mathematical problem is the Initial Value Problem (IVP).

Of course, many biological, physical, and social systems require more than one variable to define a state. Even a simple pendulum represented by a ball of mass m attached to a massless, rigid rod with a pivot point so that the ball moves in a plane requires two variables to describe the state of the system, the angle θ measured with respect to the vertical and the angular velocity ω. Figure 7.6 shows the geometry.

The more general dynamical system can be described by N state variables, $y_0, ..., y_{N-1}$, and N rate equations, one for each state variable.

$$\frac{dy_0}{dt} = f_1\left(t, y_0, \cdots, y_{N-1}, \mathbf{p}_0\right)$$

$$\vdots \qquad\qquad\qquad (7.10)$$

$$\frac{dy_{N-1}}{dt} = f_{N-1}\left(t, y_0, \cdots, y_{N-1}, \mathbf{p}_{N-1}\right)$$

Again, we assume that the functions $f_0, \cdots, f_{N-1}$ are known. If the state variables are known at a particular time, $t = 0$, for example, then the differential equations can be solved to give the state variables at subsequent times.

7.3. Numerical Solution of a Dynamical Systems Model

Solving a dynamical system model so that state variables can be predicted for some range of t values requires solving an ordinary differential equation, Equation (7.1), or a system of such equations, Equation (7.2). For some choices of the function f, or series of functions $f_0, ..., f_{N-1}$, techniques from the theory of ordinary differential equations can be used to obtain explicit solutions for y or $y_0, \ldots, y_{N-1}$. For example, consider the rate equation that leads to exponential growth.

$$\frac{dy}{dt} = ry\left(t\right) \qquad\qquad (7.11)$$

Using elementary methods from calculus, this differential equation can be integrated to obtain

$$y\left(t\right) = y\left(0\right)e^{rt} \qquad\qquad (7.12)$$

where $y(0)$ is the value of y at $t = 0$.

We will often encounter dynamical systems models for which an explicit solution for the state variables cannot be obtained, at least not by the scientist or engineer who needs to use the solution. For such models, a numerical solution can be obtained through a numerical integration of the differential equation. We will not discuss here the mathematics behind developing such techniques but will present one method for using these techniques. The mathematics behind the numerical methods we will use is discussed by Iserles (Iserles, 2009). A powerful method for doing the numerical solution is to use the solve_ivp function from the scipy.integrate library (The SciPy Community, 2022). Figure 7.7 gives the basic template for using solve_ivp for a one-state variable model, as shown in Equation 7.9.

```
1  """
2  Program Name: Exponential Growth Model: Numerical Solution
3  Author: C.D. Wentworth
4  version: 3.17.2020.1
5  Summary: Basic script for solving a dynamical system representing
6          exponential growth. A numerical integration technique is
7          used.
8
9  """
10
11 import scipy.integrate as si
12 import numpy as np
13 import matplotlib.pylab as plt
14
15 def f(t,y,r):
16 #   y = a list that contains the system state
17 #   t = the time for which the right-hand-side of the system equations
18 #       is to be calculated.
19 #   r = a parameter needed for the model
20 #
21     import numpy as np
22
23 #   Unpack the state of the system
24     y0 = y[0] # cell density
25
26 #   Calculate the rates of change (the derivatives)
27     dy0dt = r*y0
28
29     return [dy0dt]
```

Figure 7.7a: Template for using solve_ivp.

```
31  # Main Program
32
33  # Define the initial conditions
34  yi = [0.021]
35
36  # Define the time grid
37  t = np.linspace(0,300,200)
38
39  # Define the model parameters
40  r = 0.014
41  p = (r,)
42
43  # Solve the DE
44  sol = si.solve_ivp(f,(0,300),yi,t_eval=t,args=p)
45  ys = sol.y[0]
46
47  # Plot the theory
48  plt.plot(t,ys,color='g')
49  plt.xlabel('t [min]')
50  plt.ylabel('OD')
51  plt.savefig('ExponentialGrowth.png',dpi=300)
52  plt.show()
```

Figure 7.7b: Continuation of Template for using solve_ivp.

As the code shows, you must define a Python function that calculates the right hand side of the rate equation defined in Equation (7.1) in terms of t and the current value of the state variable y, symbolized by the parameter y in $f(t, y, r)$. You must also define the initial value $y(0)$, called yi in the program, and a list of t values for which you need the value of the state variable $y(t)$, which is done in line 37. The function solve_ivp returns a list of objects, contained in sol, that defines the solution. The solution corresponding to the times in the array t are contained in the array sol.y[0]. To use the solution values in a plot we extract the actual solution values from sol and name the array ys, in line 45. If the dynamical systems model had more than one state variable then we could extract the solution values for each state variable as shown below.

```
ys0 = sol.y[0]
ys1 = sol.y[1]
 .
 .
 .
```

184

We will see an example of doing this in the next chapter. If the model has more than one parameter, they must all be placed in the tuple p in line 41 but listed separately in the function definition. The next section will show an example of such a model.

7.4. Computational Problem Solving: Mathematical Model for the Growth of Pseudomonas aeruginosa

To illustrate the numerical solution of a dynamical systems model we will return to the problem presented at the beginning of the chapter: the growth of *Pseudomonas aeruginosa* in a container with a fixed amount of liquid growth medium. When our goal is to develop a mathematical model of a system, we will use a modification of our four-step computational problem-solving strategy. Our strategy is

1. Analyze the problem
2. Formulate a model
3. Solve the model
4. Verify and interpret the model's solution
5. Report on the model
6. Maintain the model

Analyze the Problem

Our analysis begins with some library research on growth of microorganisms. We find that the typical stages of growth for bacteria in a container with fixed amount of growth medium are those shown in Figure 7.8 (Hardy, 2002). This is supposed to be a semi-log plot with the y-axis being the log axis. On such a plot, exponential growth appears as a straight line.

For now, we will concentrate on modeling the exponential phase of cell growth. What factors in the system might affect the growth? In general, we might expect

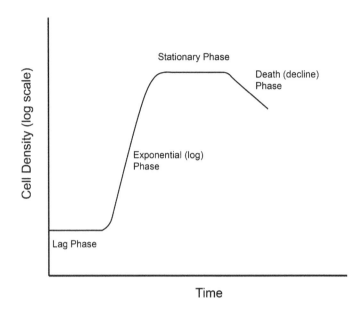

Figure 7.8.: Typical stages of growth for bacteria in a fixed amount of medium.

the temperature, pH, and other environmental factors will influence growth. We will assume those environmental factors are held constant. We are being very restrictive in how we describe the system. This is typical in starting a modeling project. We start with simple models and work our way up in complexity. In our case, there is one state variable:

$$y\,(t) \equiv \text{number of cells per ml} \tag{7.13}$$

What will determine the rate of change of $y\,(t)$? A reasonable hypothesis is that the rate of change of y will depend on the current number of cells, or cell density, that are present.

Formulate a model

We can express the hypothesis stated above with Equation 7.14, repeated below.

$$\frac{dy}{dt} = ry(t) \tag{7.14}$$

Solve the model

The model defined by Equation 7.14 can be solved with a little bit of calculus to
give

$$y(t) = y(0) e^{rt} \tag{7.15}$$

This is the exponential growth model.

We will implement a numerical solution below since it is more common for the
scientist or engineer to require such an approach.

Verify and interpret the model's solution

We have data for the state variables with which to compare the predictions from
the model. This is the ideal way of verifying the model. The data used to produce
Figure 7.1 is in the file PA01_SuspCell_0p25PerCentGlucose_5-3-2017.txt. This
data uses optical density of the liquid in the growth container to represent the
cell density. Optical density is usually proportional to the actual cell density but
establishing the exact relationship can be time-consuming.

The code in Figure 7.7 can be adapted to display both the theory and the data. Since
the data file is composed of simple numerical columns we use the loadtxt function
from Numpy to read in the data rather using Pandas. We can get the initial value of
the state variable from the first row of data in the file. The value of the growth rate,
r, is adjusted through trial and error to improve the fit between model and data.
We will use statistical methods in Chapter 10 to find the best-fit model parameters.
Figure 7.10 shows the comparison of the model and data.

```
1  """
2  Program Name: PA01 Growth - Model and Data
3  Author: C.D. Wentworth
4  version: 8.16.2022.1
5  Summary: This program performs a numerical solution for the exponential
6          growth dynamical systems model and compares the model results
7          with actual growth data for PA01.
8  History:
9      8.16.2022.1: base
10
11 """
12
13 import scipy.integrate as si
14 import numpy as np
15 import matplotlib.pylab as plt
16
17 def f(t, y, r):
18     #   y = a list that contains the system state
19     #   t = the time for which the right-hand-side of the system equations
20     #       is to be calculated.
21     #   r = model parameter (specific growth rate)
22     #
23     import numpy as np
24
25     #   Unpack the state of the system
26     y0 = y[0]   # cell density
27
28     #   Calculate the rates of change (the derivatives)
29     dy0dt = r * y0
30     return [dy0dt]
```

Figure 7.9a: Code for exponential growth model calculation and data comparison.

```python
32  # Main Program
33
34  # Read in data
35  cols = np.loadtxt('PA01_SuspCell_0p25PerCentGlucose_5-3-2017.txt',skiprows
        =6)
36  timeData = cols[:,0]
37  ODData = cols[:,1]
38
39  # Define the initial conditions
40  yi = [0.074]
41
42  # Define the time grid
43  t = np.linspace(0, 400, 200)
44
45  # Define the model parameters
46  r = 0.01175
47  p = (r,)
48
49  # Solve the DE
50  sol = si.solve_ivp(f, (0, 400), yi, t_eval=t, args=p)
51  ys = sol.y[0]
52
53  # Plot the theory and data
54  plt.plot(t, ys, color='g', label='Model')
55  plt.plot(timeData, ODData, linestyle='', marker='d',
56          markersize=5.0, color='blue', label='Data')
57  plt.xlabel('t [min]')
58  plt.ylabel('OD')
59  plt.yscale('log')
60  plt.title('PA01 Growth: Comparison with Exponential Growth Model',
61          fontsize= 12)
62  plt.legend()
63  plt.savefig('PA01_ExpGrowth_ModelAndData.png', dpi=300)
64  plt.show()
```

Figure 7.9b: Continuation of code for exponential growth model calculation and data comparison.

The exponential growth model shows a clear deficiency in describing the PA01 growth data for $t > 200[\text{min}]$. We can improve our model by including the effect of a finite amount of nutrients. At some point there will not be enough nutrients to allow cells to grow and divide. We must replace the rate equation 7.11 with one that shows exponential growth for the rate equation for small values of y but eventually has the rate go to zero. One possible hypothesis is

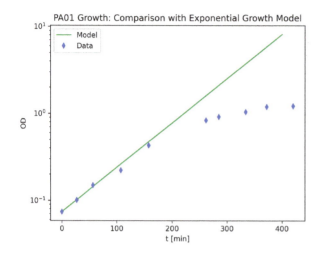

Figure 7.10.: Comparison of exponential growth model with PA01 data.

$$\frac{dy(t)}{dt} = ry\left(1 - \frac{y}{M}\right) \tag{7.16}$$

M is a new model parameter that represents the carrying capacity of the growth environment. When y approaches M, the rate of change for y will go to zero. The rate equation in Equation 7.16 is a historically famous one that defines the logistics model. Calculus techniques can be used to solve this differential equation exactly yielding Equation 7.17.

$$y(t) = \frac{My(0)e^{rt}}{M + y(0)(e^{rt} - 1)} \tag{7.17}$$

Instead of using the exact solution of Equation 7.16 we will implement a numerical solution so as to gain a little more experience in setting up the numerical solution framework.

```
1  """
2  Program Name: PA01 Growth - Logistics Model and Data
3  Author: C.D. Wentworth
4  version: 8.16.2022.1
5  Summary: This program performs a numerical solution for the logistics
6          growth dynamical systems model and compares the model results
7          with actual growth data for PA01.
8  History:
9      8.16.2022.1: base
10
11 """
12
13 import scipy.integrate as si
14 import numpy as np
15 import matplotlib.pylab as plt
16
17 def f(t, y, r, M):
18     #   y = a list that contains the system state
19     #   t = the time for which the right-hand-side of the system equations
20     #       is to be calculated.
21     #   r = the specific growth rate
22     #   M = the carrying capacity
23     #
24     import numpy as np
25
26     #   Unpack the state of the system
27     y0 = y[0]   # cell density
28
29     #   Calculate the rates of change (the derivatives)
30     dy0dt = r * y0 * (1.0 - y0/M)
31     return [dy0dt]
```

Figure 7.11a: Code for logistics growth model calculation and data comparison.

```
32  # Main Program
33
34  # Read in data
35  cols = np.loadtxt('PA01_SuspCell_0p25PerCentGlucose_5-3-2017.txt',skiprows
        =6)
36  timeData = cols[:,0]
37  ODData = cols[:,1]
38
39  # Define the initial conditions
40  yi = [0.074]
41
42  # Define the time grid
43  t = np.linspace(0, 415, 200)
44
45  # Define the model parameters
46  r = 0.014
47  M = 1.25
48  p = (r,M)
49
50  # Solve the DE
51  sol = si.solve_ivp(f, (0, 415), yi, t_eval=t, args=p)
52  ys = sol.y[0]
53
54  # Plot the theory and data
55  plt.plot(t, ys, color='g', label='Model')
56  plt.plot(timeData, ODData, linestyle='', marker='d',
57          markersize=5.0, color='blue', label='Data')
58  plt.xlabel('t [min]')
59  plt.ylabel('OD')
60  plt.yscale('log')
61  plt.title('PA01 Growth: Comparison with Logistics Growth Model',
62          fontsize= 12)
63  plt.legend()
64  plt.savefig('PA01_LogisticsGrowth_ModelAndData.png', dpi=300)
65  plt.show()
```

Figure 7.11b: Continuation of code for logistics growth model calculation and data comparison.

Report on the model

We used two different dynamical systems models to describe the PA01 bacteria growing in a glucose medium. The exponential growth model described by Equation 7.14 did not compare well with the available data over the time period in the data set, as can be seen in Figure 7.10. The logistics model given by Equation 7.16

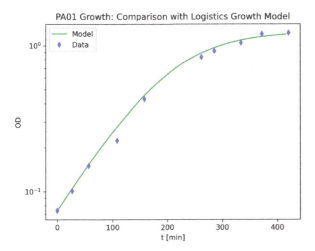

Figure 7.12.: Comparison of logistics growth model with PA01 data.

described the available data well. as seen in Figure 7.12. The model parameter
values that are appropriate for the PA01 strain of *Pseudomonas aeruginosa* grown
in a 0.25% glucose minimal medium as 37 [C] are

$$r = 0.014 \left[\text{min}^{-1}\right]$$
$$M = 1.25 \left[\text{OD units}\right]$$

(7.18)

Maintain the model

If we could obtain growth data over a longer period of time, then we would see
the stationary phase give way to the death phase. We would then need to develop
a more sophisticated model that could account for this phase. This could be the
subject of a future research project.

7.5. Exercises

1. A dynamical system is one that changes with ____.

2. Select the statements that are true for a mathematical dynamical systems model.

  a. The state of the system is described by a set of time-dependent variables.

  b. The time rate of change of each state variable must be known.

  c. The current state of the system does not depend on the past state.

  d. It is a stochastic model.

3. T/F: The solve_ivp function can solve a first-order differential equation using numerical methods.

4. Calculate the integral of $v(t)$ from t = 0 to t = 2 [s] where $v(t)$ is shown in the figure below.

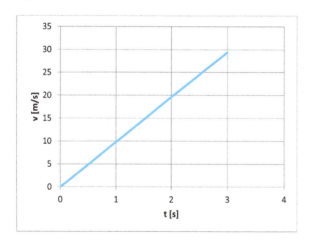

5. Write a short Python program that will calculate the derivative of the function

$$f(x) = \left(3x + 2x^3\right)\cos(x)$$

at $x = 2$. Use the cosine function in the Numpy module.

6. Write a short Python program that will calculate the integral from $0 \le x \le 4$ of the following function

$$f\left(x\right) = \left(3x + 2x^3\right)\cos\left(x\right)$$

7.6. Program Modification Problems

1. Your goal is to use a numerical solution to an exponential growth model to simulate bacteria growth in an unconstrained environment (plenty of food) and find the growth rate parameter for *Escherichia coli* grown in CSHA media. Start with the code in Ch7ProgModProb1.py . Add a section that reads in the growth data in BacterialGrowthData.txt. Make sure you read in the column with the CSHA optical density measurement. Calculate the exponential growth model values for the optical density over the same time period as the data. Plot the data on the same graph. Change the value of the growth rate parameter and make the theory fit the data as best as you can.

Your plot should have the following features:

■ The theoretical (exponential growth model) values should be plotted with a solid green line.

■ The data points should be plotted using a filled circle as the marker. Make the color red and the size 8.

■ Add an appropriate title.

■ Save the graph as a png file.

2. Your goal is to use a numerical solution to a logistics growth model to simulate bacteria growth in a constrained environment (limitation in food) and find the growth rate parameter and carrying capacity for *Vibrio natriegens* grown in TSB

media. Start with the code in Ch7ProgModProb2.py .

Change the import data section that reads in the growth data in V_natriegens-GrowthData.txt. Make sure you read in the column with the TSB optical density measurement.

Change the function $f(t, y, r)$ to implement the logistics model rate equation.

Change the initial condition.

Change the model parameter tuple p so that it contains both the growth rate r and the carrying capacity M.

Change the value of the growth rate parameter and carrying capacity and make the theory fit the data as best as you can.

Create a plot of both theory and data on the same graph. Your plot should have the following features:

❙ The theoretical (logistics model) values should be plotted with a solid green line.

❙ The data points should be plotted using a triangle as the marker. Make the color red and the size 8.

❙ Save the graph as a png file.

7.7. Program Development Problems

1. Develop an argument that the numerical solution of a dynamical system model that we are using is valid. This will involve choosing a specific model for which we know the exact solution. The exponential growth model would be one choice. Next, create a program that solves the model numerically, add a function that calculates the exact solution, and then create a table that compares the numerical and the exact values for 10 values of the independent variable. Write a brief essay that summarizes your evidence that the numerical solution method is valid.

2. Solve the following first-order differential equation with the given initial condition by developing a Python program that can solve the equation numerically.

$$\frac{dy(t)}{dt} = \frac{(e^{rt} - 2y(t))}{(1 + y^2(t))}, y(0) = 0$$

Your program should make a plot of the solution for $0 \leq t \leq 5$ and for $r_1 = 0.1$ and $r_2 = 0.5$. The plot should have a solid blue continuous line for the model solution with r_1 and a solid red line for the model solution with r_2. Add appropriate axis titles, a legend, and a chart title (y(t) versus t).

7.8. References

Diggle, S. P., & Whiteley, M. (2020). Microbe Profile: Pseudomonas aeruginosa: opportunistic pathogen and lab rat. *Microbiology, 166*(1), 30–33. https://doi.org/10.1099/mic.0.000860

Hardy, S. P. (2002). Chapter 2: Bacterial Growth. In *Human Microbiology.* Taylor & Francis.

Iserles, A. (2009). *A first course in the numerical analysis of differential equations* (2nd ed., Vol. 1–1 online resource (xviii, 459 pages) : illustrations). Cambridge University Press; WorldCat.org. http://www.books24x7.com/marc.asp?bookid=30922

The SciPy community. (2022a). *scipy.integrate.quad—SciPy v1.9.0 Manual.* https://docs.scipy.org/doc/scipy/reference/generated/scipy.integrate.quad.html

The SciPy Community. (2022). *scipy.integrate.solve_ivp—SciPy v1.9.0 Manual.* https://docs.scipy.org/doc/scipy/reference/generated/scipy.integrate.solve_-ivp.html

The SciPy community. (2022b). *scipy.misc.derivative—SciPy v1.9.0 Manual.* https://docs.scipy.org/doc/scipy/reference/generated/scipy.misc.derivative.html

8. Dynamical Systems Modeling II

Chapter 7 introduced the concept of a dynamical system, a type of deterministic mathematical model with behavior determined by a first-order differential equation, the rate equation for the state variable. This chapter will expand the type of model that we can study to include more than one state variable. One consequence of this generalization is that we will develop the capability to solve problems based on a Newton's second law analysis of a system. This will open up significant areas of physics and engineering to us.

Motivating Problem: The Zombie Apocalypse

Zombies are a fixture of modern popular culture. Wikipedia currently lists 588 movies with a zombie theme (Wikipedia Contributors, 2022). If we add in all of the television series, comics, novels, short stories, and electronic games, the list of zombie-themed entertainment titles would undoubtedly be in the thousands. Even the Center for Disease Control offers a pamphlet on preparing for a zombie pandemic (Centers for Disease Control and Prevention (U.S.) & Office of Public Health Preparedness and Response, 2011).

The first Hollywood box-office hit with a zombie theme was White Zombie, released in 1932 and starring Bela Lugosi, of Dracula fame, as Voodou master "Murder" Legendre, who kills and resurrects people as zombies using a potion (Kay & Brugués, 2012). The zombies in this film are mindless beasts controlled by their maker, not the brain-eating, decayed flesh monstrosities of more modern productions.

The 1968 film Night of the Living Dead defined critical elements of the modern zombie genre: normal humans become infected from the bite of a zombie, an infected person dies and becomes reanimated with an appetite for human flesh, and a zombie can be killed with a shot to the base of the skull or by fire. Night of the Living Dead did not use the term zombie for the reanimated, flesh-eating wanderers. These creatures were called ghouls in the movie.

The modern zombie story views the condition as an infectious disease spread through close contact with uninfected people. Typically, the infection is incredibly successful at spreading through the human population, leaving the world with a huge number of roaming, flesh-eating creatures in the midst of a dwindling number of humans. Figure 8.1 shows one depiction of such a pandemic, as visualized by an artist at the U.S. Centers for Disease Control (Centers for Disease Control and Prevention (U.S.) & Office of Public Health Preparedness and Response, 2011). Our modeling problem is to understand the conditions for which the zombie infection can spread so as to essentially wipe out the human population. A related question is, what are the conditions that will allow humans to make a comeback?

Figure 8.1.: The zombie pandemic, as pictured by the CDC (Centers for Disease Control and Prevention (U.S.) & Office of Public Health Preparedness and Response, 2011).

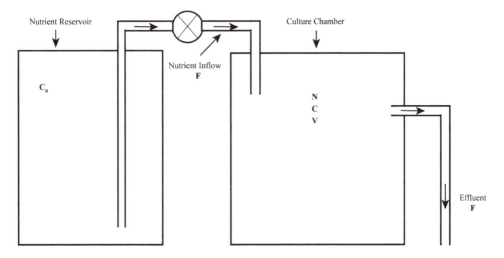

Figure 8.2.: Physical structure of a simple chemostat.

8.1. Compartmental Models

Compartmental models are a class of mathematical model for which the system under investigation is assumed to be composed of populations that can exist in one or more states or compartments. Each compartment can be described by one state variable and flows can occur between the compartments. In addition to flows between compartments, other processes that affect the population in a compartment can be included, such as birth and death events.

Compartmental models have become useful in many areas of science and engineering including chemical engineering, ecology, environmental engineering, epidemiology, pharmacokinetics, and physiology. To introduce the concept, we will focus on developing a compartmental model of a chemostat, a type of bioreactor used in microbiology research and in bioengineering applications.

A chemostat provides a continuous supply of microorganisms using constant and reproducible growth conditions. This allows the microorganisms to be used in research or industrial applications under predictable and controllable conditions. Figure 8.2 shows the conceptual physical structure of a chemostat.

Table 8.1.: Chemostat model variables.

Variable	Symbol	Units	Units Example
Concentration of bacteria in the culture chamber	N	cells/volume	cells/ml
Nutrient concentration in the culture chamber	C	mass/volume	g/ml
Nutrient concentration in the nutrient reservoir	C0	mass/volume	g/ml
Volume of the culture chamber	V	volume	ml
Nutrient inflow and effluent outflow rate	F	volume/time	ml/s

We can start constructing a mathematical model of this device defining important variables or parameters that describe the system.

We will make some assumptions about our system to keep our model relatively simple.

1. The nutrient reservoir will always supply nutrient at the fixed concentration C_0.
2. The nutrient inflow rate will always equal the outgoing effluent rate, F. This will keep the volume of the culture chamber constant.
3. The culture chamber is well mixed, so the bacteria and nutrient concentrations are constant throughout.

With these assumptions we can focus our attention on the culture chamber containing both the bacteria and the nutrient. Our system can be modeled conceptually by two compartments: a bacteria compartment and a nutrient compartment. Our conceptual compartmental model of the system is represented in Figure 8.3.

To define the dynamical system model we must specify rate equations for the two state variables, N and C. This will require some additional model parameters, defined in Table 8.2.

The rate equation for a state variable must contain terms corresponding to each

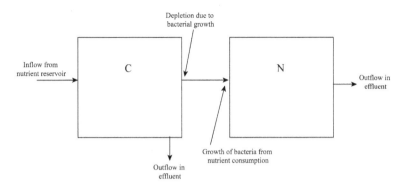

Figure 8.3.: Compartmental model of the system.

Table 8.2.: Chemostat model parameters.

Parameter	Symbol	Units
the yield constant	Y	(cells/volume)/(mass/volume)
specific growth rate	K	1/time
maximum specific growth rate	K_{max}	1/time
saturation constant	K_s	mass/volume

arrow going into or out from the associated compartment in Figure 8.3. Following Leah Edelstein-Keshet, we will hypothesize the following rate equations (Edelstein-Keshet, 2005).

$$\frac{dN}{dt} = K\left(C\right)N - \frac{FN}{V} \tag{8.1}$$

$$\frac{dC}{dt} = -\frac{K\left(C\right)N}{Y} - \frac{FC}{V} + \frac{FC_0}{V} \tag{8.2}$$

The specific growth rate, K, is assumed to be a function of the nutrient concentration, C. The yield constant, Y, maximum specific growth rate, K_{max}, and the saturation constant, K_s, will depend on the species of bacteria being modeled.

We must also hypothesize a model for the specific growth rate. A realistic model should show the growth rate reaching a maximum for large values of the nutrient concentration. A commonly used model is described by Michaelis-Menten kinetics equation (Edelstein-Keshet, 2005):

$$K\left(C\right) = \frac{K_{\max}C}{K_s + C} \tag{8.3}$$

With this choice for the growth kinetics model, the rate equations become

$$\frac{dN}{dt} = \left(\frac{K_{\max}C}{K_s + C}\right)N - \frac{FN}{V} \tag{8.4}$$

$$\frac{dC}{dt} = -\left(\frac{K_{\max}C}{K_s + C}\right)\frac{N}{Y} - \frac{FC}{V} + \frac{FC_0}{V} \tag{8.5}$$

This model has six parameters: $K_{\max}$, K_s, Y, V, F, C_0. Exploring the behavior of the model can be complicated with so many parameters. Another complication that arises in performing numerical solutions of the model is that there can be a large range of numbers involved. Both of these complications can be mitigated to some extent by rewriting the equations in a dimensionless form. This can be done by writing the original variables N, C, and t as the product of a dimensionless variable and a dimensional number (Edelstein-Keshet, 2005).

$$N = N^* \times \widehat{N}$$
$$C = C^* \times \widehat{C} \tag{8.6}$$
$$t = t^* \times \widehat{t}$$

In these definitions, N^*, C^*, and t^* are the new dimensionless variables and $\widehat{N}, \widehat{C}$, and $\widehat{t}$ are constants that contain dimensions. Our goal is to choose the values of these dimension-carrying constants in a way that reduces the number of model parameters in Equations (8.3) and (8.4). Substituting the definitions from Equation (8.5) into Equations (8.3) and (8.4) yields Equations (8.6) and (8.7).

$$\frac{dN^*}{dt^*} = \hat{t}K_{\max}\left(\frac{C^*}{K_s/\hat{C} + C^*}\right)N^* - \hat{t}\frac{FN^*}{V} \tag{8.7}$$

$$\frac{dC^*}{dt^*} = -\left(\frac{\hat{t}K_{\max}\hat{N}}{Y\hat{C}}\right)\left(\frac{C^*}{K_s/\hat{C} + C^*}\right)N^* - \hat{t}\frac{FC^*}{V} + \hat{t}\frac{FC_0}{V\hat{C}} \tag{8.8}$$

By choosing the dimensional constants $\hat{N}, \hat{C},$ and $\hat{t}$ appropriately, the number of independent model parameters can be reduced. The following choice yields two model parameters.

$$\hat{t} = \frac{V}{F}, \hat{C} = K_s, \hat{N} = \frac{YK_s}{\hat{t}K_{\max}} \tag{8.9}$$

To facilitate performing a numerical solution of these equations to give N* and C* as functions of time, we will change the notation to the generic dynamical systems model notation.

$$N^* \rightarrow y_0$$
$$C^* \rightarrow y_1 \tag{8.10}$$

Making the substitutions defined in Equation (8.8) and the notation change defined by Equation (8.9) gives the final form of our dynamical systems model.

$$\frac{dy_0}{dt} = a_1\left(\frac{y_1}{1 + y_1}\right)y_0 - y_0 \tag{8.11}$$

$$\frac{dy_1}{dt} = -\left(\frac{y_1}{1 + y_1}\right) y_0 - y_1 + a_2 \tag{8.12}$$

where

$$a_1 = \widehat{t} K_{\max} = \frac{V K_{\max}}{F} \tag{8.13}$$

$$a_2 = \frac{\widehat{t} F C_0}{V \widehat{C}} = \frac{C_0}{K_s}$$

The model parameters a_1 and a_2 are dimensionless.

The dynamical systems model defined by Equations (8.11) and (8.12) can be solved numerically by using the code template from Figure 7.7. Figure 8.4 shows the Figure 7.7 code adapted for the chemostat problem. Figure 8.5 shows the result of running the code. For these model parameter values, it takes about 40 dimensionless time units to achieve a stable steady-state production of bacteria.

8.2. Numerical Solution of 2nd-Order Differential Equations

Many scientific and engineering systems can be modeled with second-order differential equations. One reason this is the case is that Newton's second law is really a second-order differential equation.

$$F_{net} = m a_x = m \frac{d^2 x}{dt^2} \tag{8.14}$$

We can solve such an equation numerically without being expert mathematicians by using our ability to solve a first-order equation numerically, as done in our dynamical systems models. We can make use of the dynamical systems approach to numerical solution by using a nice trick to turn a second-order differential equation into two first-order differential equations.

```python
1  """
2  Program: Chemostat Model
3  Author: C.D. Wentworth
4  Version: 8.22.2022.1
5  Summary:
6      This program implements a dynamical systems model of a
7      chemostat that uses one growth-limiting nutrient and
8      produces one species of bacteria. The model is expressed in
9      a dimensionless form.
10 Version History:
11     8.22.2022.1: base
12
13 """
14 import scipy.integrate as si
15 import numpy as np
16 import matplotlib.pylab as plt
17
18 # Create a function that defines the rhs of the differential equation
       system
19
20 def f(t, y, a1, a2):
21     #   y = a list that contains the system state
22     #   t = the time for which the right-hand-side of the system equations
23     #       is to be calculated.
24     #   a1 = a parameter needed for the model
25     #   a2 = a paramteter needed for the model
26
27     #   Unpack the state of the system
28     y0 = y[0]   # N*
29     y1 = y[1]   # C*
30
31     #   Calculate the rates of change (the derivatives)
32     dy0dt = a1*(y1/(1.0 + y1))*y0 - y0
33     dy1dt = -(y1/(1.0 + y1))*y0 - y1 + a2
34
35     return [dy0dt, dy1dt]
```

Figure 8.4a: Adaptation of the dynamical systems code for the chemostat problem.

```
38  # Main Program
39
40  # Define the initial conditions
41  yi = [0.01, 0.8]
42
43  # Define the time grid
44  ti = 0.0
45  tf = 60.0
46  t = np.linspace(ti, tf, 100)
47
48  # Define the parameter tuple
49  a1 = 2.6   #
50  a2 = 0.8   #
51  p = (a1, a2)
52
53  # Solve the DE
54  sol = si.solve_ivp(f, (ti, tf), yi, t_eval=t, args=p)
55  N = sol.y[0]
56  C = sol.y[1]
57
58  # Plot the solution
59  plt.plot(t, N, color='green', label='N*')
60  plt.plot(t, C, color='red', label='C*')
61  plt.xlabel('t*')
62  plt.ylabel('N* , C*')
63  plt.legend()
64  plt.savefig('chemostat.png')
65  plt.show()
```

Figure 8.4b: Continuation of the adaptation of the dynamical systems code for the chemostat problem.

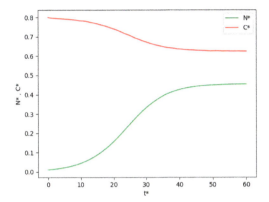

Figure 8.5.: The dimensionless cell density and nutrient concentration as functions of time.

The key will be to define a new set of variables that produce a new model that is simply related to the old second-order differential equation model. We will learn how to define the new variables by considering a specific example. Equation 8.15 defines a second-order differential equation.

$$\frac{d^2 y(t)}{dt^2} + q(t) \frac{dy(t)}{dt} = r(t) \tag{8.15}$$

where we assume that the functions $q(t)$ and $r(t)$ are known. Define a new set of variables $y_0(t)$ and $y_1(t)$ as follows

$$y_0 = y$$
$$y_1 = \frac{dy}{dt} \tag{8.16}$$

The rate of change of $y_1(t)$ can be obtained from the original second-order differential equation:

$$\frac{d^2y\,(t)}{dt^2} = r\,(t) - q\,(t)\,\frac{dy\,(t)}{dt} \Rightarrow \frac{dy_1\,(t)}{dt} = r\,(t) - q\,(t)\,y_1\,(t) \qquad (8.17)$$

This is due to the second derivative of y being equal to the first derivative of y_1. So, Equation 8.15 can be replaced by the following set of first-order differential equations:

$$\frac{dy_0\,(t)}{dt} = y_1\,(t)$$

$$\frac{dy_1\,(t)}{dt} = r - qy_1\,(t) \qquad (8.18)$$

If we solve this system numerically, then we will have also solved Equation 8.15 for $y(t)$, since that is just the new variable y_0.

8.3. Newton's 2nd Law Problems

Let's look at performing a numerical solution to a specific problem: a mass attached to a spring. This is a good testing problem since it can be solved exactly without performing the numerical solution. This will allow us to verify our numerical approach.

Figure 8.6 shows the scenario: a block of mass 0.15 [kg] is attached to a spring with spring constant of 20 [N/m]. The mass is stretched a distance $x_0 = 0.10$ [m] and released from rest. The state variable x will be the position of the mass measured from the spring's unstretched position.

We can create a model of the object's motion by using Newton's second law and Hooke's law for the spring force, which is the only force in the x direction.

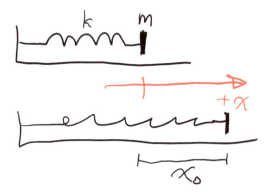

Figure 8.6.: . Mass attached to a spring and given an initial stretch.

$$F_{net} = m\frac{d^2x}{dt^2} \tag{8.19}$$

$$F_s = -kx \tag{8.20}$$

Combining these equations leads to our model for x:

$$\frac{d^2x\,(t)}{dt^2} = -\frac{k}{m}x\,(t) \tag{8.21}$$

Now we apply our trick to turn the second-order differential equation into two first-order equations. We define new state variables $y_0\,(t)$ and $y_1\,(t)$:

$$\begin{aligned} y_0 &= x \\ y_1 &= \frac{dy_0}{dt} \end{aligned} \tag{8.22}$$

The rate equations for our new state variables are

$$\frac{dy_0}{dt} = y_1$$

$$\frac{dy_1}{dt} = -\frac{k}{m}y_0 = -\frac{20[N/m]}{0.15[kg]}y_0$$

(8.23)

with the initial conditions

$$y_0(0) = x_0 = 0.10\,[m]\ ,\ y_1(0) = 0$$

(8.24)

The model is now in a form that can be solved numerically using our Dynamical Systems Template adapted to this model. Figure 8.7 shows the code that will solve this model numerically. Figure 8.8 shows the graphical output produced by the code.

8.4. Computational Problem Solving: Modeling the Zombie Apocalypse

We now return to the problem of modeling the Zombie Apocalypse introduced at the beginning of the chapter.

8.4.1. Analysis

The zombie problem we will model is one where it is an infectious disease that requires contact between a susceptible person and an undead individual. We will use a recently published academic study of this type of zombie by infectious disease scientists to help us develop a model (Munz et al., 2009). The population of humans and zombies is assumed to be composed of three compartments:

■ Susceptibles (S)

```python
1  """
2  Program: Ideal Spring Simulation
3  Author: C.D. Wentworth
4  Version: 2.20.2022.1
5  Summary:
6      This program implements a dynamical systems model of the
7      one-dimensional ideal spring (Hooke's Law).
8  Version History:
9      2.20.2022.1: base
10
11
12  """
13
14  import scipy.integrate as si
15  import numpy as np
16  import matplotlib.pylab as plt
17
18  # Create a function that defines the rhs of the differential equation
       system
19
20  def f(t, y, m, k):
21      #   y = a list that contains the system state
22      #   t = the time for which the right-hand-side of the system equations
23      #       is to be calculated.
24      #   k = a parameter needed for the model - force constant
25      #   m = a paramteter needed for the model - mass
26
27  #   Unpack the state of the system
28      y0 = y[0]   # x
29      y1 = y[1]   # vx
30
31  #   Calculate the rates of change (the derivatives)
32      dy0dt = y1
33      dy1dt = -(k/m)*y0
34
35      return [dy0dt, dy1dt]
```

Figure 8.7a: Python code for solving the spring problem using the dynamical system numerical solution method.

```
37 # Main Program
38
39 # Define the initial conditions
40 yi = [0.10, 0.0]
41
42 # Define the time grid
43 ti = 0.0
44 tf = 2.0
45 t = np.linspace(ti, tf, 100)
46
47 # Define the parameter tuple
48 m = 0.15   # mass in [kg]
49 k = 20.0   # force constant in [N/m]
50 p = (m, k)
51
52
53 # Solve the DE
54 sol = si.solve_ivp(f, (ti, tf), yi, t_eval=t, args=p)
55 x = sol.y[0]
56 vx = sol.y[1]
57
58
59 # Plot the solution
60 plt.plot(t, x, color='green', label='x')
61 plt.xlabel('t [s]')
62 plt.ylabel('x [m]')
63 plt.savefig('idealSpring.png')
64 plt.show()
```

Figure 8.7b: Continuation of the Python code for solving the spring problem using the dynamical system numerical solution method.

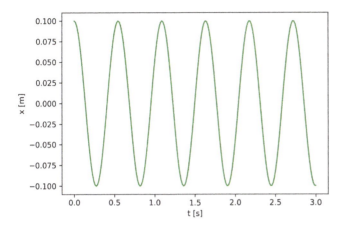

Figure 8.8.: Graph produced by the spring program.

■ Zombie (Z)

■ Removed (R)

Assumptions we will make include

1. The susceptible population can die of natural causes.
2. The susceptible population can increase due to natural growth (birth).
3. The removed compartment contains humans who have died from natural causes and from fatal zombie encounters.
4. Members of the removed compartment can resurrect as zombies.
5. The zombie compartment can increase due to non-fatal susceptible-zombie interactions (infection) or from resurrection from the removed compartment.

Figure 8.9 shows the compartmental model hypothesized by Munz, et al. (Munz et al., 2009). Table 8.3 lists the state variables and parameters used in the model.

8.4.2. Design

Defining the dynamical systems model for our zombie infection requires hypothesizing rate equations for each of the state variables. The compartmental model

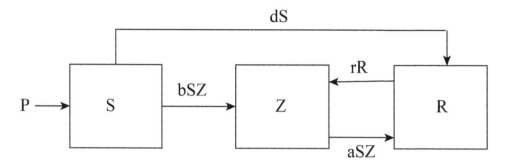

Figure 8.9.: Compartmental model for the zombie disease.

Table 8.3.: Variables and model parameters for the SZR model.

Variable or Model Parameter	Symbol
Susceptible Population	S
Zombie Population	Z
Removed Population	R
Constant Birth Rate for Susceptibles	P
Natural Death rate for Susceptibles	d
Rate at which Susceptible-Zombie encounters produce Zombies	b
Rate at which Susceptible-Zombie encounters produce dead Zombies	a
Rate at which Removed individuals resurrect as Zombies	r

diagram shown in Figure 8.9 suggest the following rate equations.

$$\frac{dS}{dt} = P - bSZ - dS \tag{8.25}$$

$$\frac{dZ}{dt} = bSZ + rR - aSZ \tag{8.26}$$

$$\frac{dR}{dt} = dS + aSZ - rR \tag{8.27}$$

We can use the dynamical systems template from Figure 7.7 to perform a numerical solution of this system. The code implementation will be facilitated by making the following notation change.

$$\begin{aligned} S &\rightarrow y_0 \\ Z &\rightarrow y_1 \\ R &\rightarrow y_2 \end{aligned} \tag{8.28}$$

The rate equations 8.25 – 8.27 become

$$\frac{dy_0}{dt} = P - by_0y_1 - dy_0 \tag{8.29}$$

$$\frac{dy_1}{dt} = by_0y_1 + ry_2 - ay_0y_1 \tag{8.30}$$

$$\frac{dy_2}{dt} = dy_0 + ay_0y_1 - ry_2 \tag{8.31}$$

8.4.3. Implementation

Performing a numerical solution for the model described by Equations 8.29 – 8.31 is straightforward. We must add one more rate calculation in the Python f function and pass the required model parameters. The revised code is in Figure 8.10 and in the code file SZRModel.py. The units for the compartment populations is arbitrary. The initial conditions must specify a nonzero number for the zombie compartment for there to be any infection activity.

```python
"""
Program: SZR Model
Author: C.D. Wentworth
Version: 8.28.2022.1
Summary:
    This program implements a dynamical systems model of a
    zombie infection based on a three compartment model.
Version History:
    8.28.2022.1: base

"""
import scipy.integrate as si
import numpy as np
import matplotlib.pyplot as plt

# Create a function that defines the rhs of the differential equation
    system

def f(t, y, P, b, d, a, r):
    #   y = a list that contains the system state
    #   t = the time for which the right-hand-side of the system equations
    #       is to be calculated.
    #   a1 = a parameter needed for the model
    #   a2 = a paramteter needed for the model

    #   Unpack the state of the system
    y0 = y[0]   # S
    y1 = y[1]   # Z
    y2 = y[2]   # R

    #   Calculate the rates of change (the derivatives)
    dy0dt = P - b*y0*y1 - d*y0
    dy1dt = b*y0*y1 + r*y2 - a*y0*y1
    dy2dt = d*y0 + a*y0*y1 - r*y2

    return [dy0dt, dy1dt, dy2dt]
#
```

Figure 8.10a: Code for the SZR model.

```
38  # Main Program
39
40  # Define the initial conditions
41  yi = [100, 0.1, 0.0]
42
43  # Define the time grid
44  ti = 0.0
45  tf = 30.0
46  t = np.linspace(ti, tf, 100)
47
48  # Define the parameter tuple
49  P = 0.1
50  b = 9.5e-3
51  d = 1.e-4
52  a = 5e-3
53  r = 1.e-4
54  p = (P, b, d, a, r)
55
56  # Solve the DE
57  sol = si.solve_ivp(f, (ti, tf), yi, t_eval=t, args=p)
58  S = sol.y[0]
59  Z = sol.y[1]
60  R = sol.y[2]
61
62  # Plot the solution
63  plt.plot(t, S, color='green', label='S')
64  plt.plot(t, Z, color='red', linewidth=3, label='Z')
65  plt.plot(t, R, color='blue', linestyle='--', label='R')
66  plt.xlabel('t')
67  plt.ylabel('S , Z , R')
68  plt.legend()
69  plt.savefig('SZRModel.png')
70  plt.show()
```

Figure 8.10b: Continuation of the code for the SZR model.

8.4.4. Testing

When the code in Figure 8.10 is executed with the model parameter choices given in lines 49-53 and with the initial conditions in line 41, the result is shown in Figure 8.11. We see that the susceptible population (uninfected humans) eventually goes to a small number close to zero and the number of zombies and removed become constant. This behavior is consistent with many zombie movies and television shows, which tend to show zombies wiping out most of the human population. This provides some evidence that our model and numerical implementation of its solution are correct.

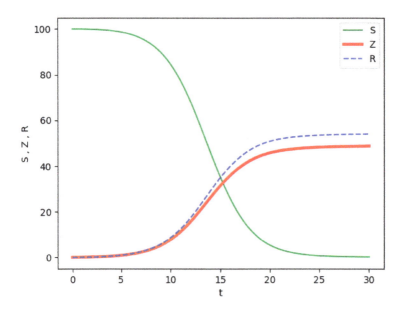

Figure 8.11.: Result of running the SZR model code.

8.5. Exercises

1. True/False: The compartments that are used in a compartmental model must correspond to actual physical containers.

2. True/False: A Newton's second law problem cannot be solved numerically using the dynamical systems template developed in Chapter 7 since Newton's second law yields a second-order differential equation rather than just a first-order equation. (Explain your answer.)

3. Replace the following second-order differential equation with two first-order equations.

$$\frac{d^2x}{dt^2} + at^2\frac{dx}{dt} = e^{-bt}$$

4. A chemostat is used to provide a continuous supply of *Pseudomonas aeruginosa* bacteria for a research project. The following parameters are used for modeling the chemostat (Robinson et al., 1984):

$$K_{max} = 0.40 \left[\text{h}^{-1}\right]$$
$$K_s = 2.0 \times 10^{-3} \left[\frac{mg}{mL}\right]$$
$$Y = 0.30 \left[\frac{\text{mg cell}}{\text{mg Gl}}\right]$$
$$V = 500. \left[\text{mL}\right]$$
$$F = 75 \left[\frac{\text{mL}}{\text{h}}\right]$$
$$C_0 = 1.0 \times 10^{-2} \left[\frac{mg}{mL}\right]$$

Glucose is used as the nutrient in this reactor. When the chemostat achieves steady-state, the dimensionless state variable values are

$$N* = 11.7 \,, C* = 0.599$$

Using the dimensional constants defined by Equation 8.9, find the actual steady-state values for the cell concentration, in (mg/mL), and the glucose concentration, in (mg/mL), in the chemostat vessel.

8.6. Program Modification Problems

1. Consider a large ball dropped from rest from a height. It will experience the force of gravity and an air drag force as it falls. The net force on the ball will be

$$F_{net} = -mg + kv_y^2$$

You need to develop a model for the ball's motion as it drops. After developing the equations describing the model, create a Python program that solves for the position of the ball as it drops. Use the following data:

m = 0.220 [kg]

k = 3.7×10^{-2} [Ns^2/m^2]

y(0) = 12 [m]

The program should create a graph of the ball's position up to the time that it hits the ground. Also, include a calculation of the position as a function of time for the constant acceleration case (no air drag). You can start with the code in Figure 8.7, which is in the code file idealSpring.py.

2. Consider the chemostat model in its dimensionless form, Equations 8.11 – 8.12. The parameter a_1 is easily changed by the person running the reactor by adjusting the flow rate, F. Using the chemostat code in chemostat_dimensionless.py, explore

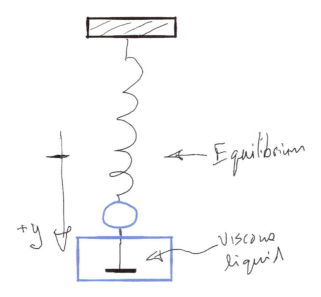

Figure 8.12.: Damped oscillator set-up.

how changing the a1 parameter affects the steady-state concentration of bacteria, N^*, and the time it takes to achieve the steady-state. Note that if a1 gets too low, corresponding to a high flow rate, the bacteria concentration goes down to a number close to zero, indicating the bacteria is getting washed out of the reactor.

Create a table of a_1, steady-state value of N^*, and time to steady-state columns. Put this table in a text file. Write a short program to create a plot of the steady-state value of N^* versus a_1. Write another program to create a plot of time to steady-state versus a_1.

8.7. Program Development Problems

1. Develop a dynamical systems model for an object attached to a spring that also experiences a damping force from motion in a viscous liquid. The figure below shows the scenario.

Assume that the spring force is given by Hooke's law.

$$F_s = -ky\,(t)$$

Assume that the damping force is proportional to the object's velocity.

$$F_d = -cv_y$$

You can ignore the gravitational force for this analysis. Use the following values for the model parameters.

$$m = 0.\,150\,[\text{kg}]$$
$$k = 40.\,0\,[\text{N/m}]$$
$$c = 1.\,50\,[\text{Ns/m}]$$

For the initial conditions, assume that the spring is stretched 15 [cm] and released from rest. This gives

$$y\,(0) = 0.15\,[\text{m}]$$
$$v_y\,(0) = 0.0\,[\text{m/s}]$$

You should use Newton's second law to find the differential equation for the object's acceleration. Convert the 2nd order differential equation into two first-order equations. Perform a numerical solution of the two-state variable model. Create a plot of the object's position as a function of time for $0 \le t \le 1.5$ [s]. Properly label the graph.

2. The SZR model introduced in Chapter 8 for modeling the zombie infection suggested that humans would always lose the battle and essentially become extinct.

Zombie stories often will offer a little more hope for human survival. Can we create another model that might include this possibility? Munz, et al. say yes. They introduce another compartment, the infected population, I, and also allow zombies can be cured of their disease. They hypothesize the following rate equations for the four compartments:

$$\frac{dS}{dt} = P - bSZ - dS + cZ$$

$$\frac{dI}{dt} = bSZ - \rho I - dI$$

$$\frac{dZ}{dt} = \rho I + rR - aSZ - cZ$$

$$\frac{dR}{dt} = dS + dI + aSZ - rR$$

Develop a program that performs a numerical solution to this revise model, which we will call the SIZR model. Since the model has seven parameters (P, b, d, c, rho, r, c), exploring its behavior is complicated. Start with a set of model parameters that are close to those used in the SZRmodel.py code shown in Figure 8.10. Assume that there is no cure so that c=0. The initial set of model parameters are

$P = 0.1$, $b = 9.5e - 3$, $d = 1.e - 4$, $c = 0$, $\rho = 3.0e - 1$, $a = 5.0e - 3$, $r = 1.0e - 4$

The initial conditions are

$$S(0) = 100 , I(0) = 0 , Z(0) = 0.1 , R(0) = 0$$

You should find that humans, the S compartment, still get wiped out, but introducing the I compartment delays the time required. Next, explore the effect of a nonzero c parameter value. A good range to explore is $0 \le c \le 0.015$. Calculate the compartment populations for $0 \le t \le 100$.

Write an essay that describes the four steps in our problem-solving strategy. In the Analysis section, provide a figure that illustrates the relationships between the four compartments implied by the rate equations given above.

8.8. References

Centers for Disease Control and Prevention (U.S.) & Office of Public Health Preparedness and Response. (2011). *Preparedness 101; zombie pandemic.* https://stacks.cdc.gov/view/cdc/6023

Edelstein-Keshet, L. (2005). *Mathematical Models in Biology.* Society for Industrial and Applied Mathematics. https://doi.org/10.1137/1.9780898719147

Kay, G., & Brugués, A. (2012). *Zombie Movies: The Ultimate Guide* (Second Edition, Second edition). Chicago Review Press.

Munz, P., Hudea, I., Imad, J., & Smith, H. L. (2009). WHEN ZOMBIES ATTACK!: MATHEMATICAL MODELLING OF AN OUTBREAK OF ZOMBIE INFECTION. In J. M. Tchuenche & C. Chiyaka (Eds.), *Infectious Disease Modelling Research Progress* (pp. 133–150).

Robinson, J. A., Trulear, M. G., & Characklis, W. G. (1984). Cellular reporoduction and extracellular polymer formation by Pseudomonas aeruginosa in continuous culture. *Biotechnology and Bioengineering, 26*(12), 1409–1417. https://doi.org/10.1002/bit.260261203

Wikipedia Contributors. (2022). List of zombie films. In *Wikipedia.* https://en.wikipedia.org/w/index.php?title=List_of_zombie_-films&oldid=1104193295

9. Stochastic Models and Simulations

One way to categorize mathematical models of natural or engineered systems is to divide them into deterministic models or stochastic models. Deterministic models, such as the dynamical systems models we have considered previously, have no element of chance in their development. The exponential growth model for bacteria is an example. If we know how many bacteria are present at $t = 0$ and the growth rate r, then we can predict the number of bacteria precisely for $t > 0$. Using Newton's Second Law to predict the trajectory of a ball given the physical forces on it is another example. There are no probabilities or uncertainties about our theoretical predications based on these models. Stochastic models, involve an element of probability or randomness in their predictions. Here is a more formal definition.

Stochastic or **probabilistic** models use random variables to describe the system. Values for the random variables are based on theoretical or empirical probability distributions.

One example of a system that is best modeled using a stochastic model is the diffusive motion of individual atoms in a gas, liquid, or solid material for which there will be a probabilistic element. Other examples would include a dynamical systems model for which the model parameters are not known precisely but take on values with some probability.

If we want to simulate the behavior of a stochastic model, then we must perform numerical experiments that use random numbers selected according to an appropriate probability distribution. Such simulations are called **Monte Carlo simulations**, which we will explore in section 9.1. Random walks are a specific kind of Monte Carlo simulation that can be useful for modeling many systems. Another category of model we will explore in this chapter is cellular automata models. This class of models is not always stochastic in nature, but often is, so we include it here.

Figure 9.1.: Roulette wheel in a casino for gambling entertainment. Photo by Derek Lynn on Unsplash.

Motivating Problem: Comparing Betting Strategies at the Roulette Wheel

Let's suppose you visit Las Vegas and want to play roulette. What kind of betting strategy should you employ so that you do not go broke? We can explore different strategies by using a computer simulation of a roulette wheel game. We want to develop a simulation for the American Roulette wheel shown in Figure 9.1 (Derek Lynn, 2020). Roulette involves a long list of possible bets ("Roulette," 2022), so we will use just a small subset of these and then explore some betting strategies such as the Martingale strategy or Anti-Martingale strategy.

9.1. Overview of Monte Carlo Methods

9.1.1. Definition

Stochastic models involve random variables. The behavior of the model will change depending on the specific values taken by the random variables in the model. Monte Carlo methods involve sampling the possible model behaviors from the set of all possible behaviors, which might by infinite, and then performing a statistical description of the behavior based on the set of samples. More succinctly, a **Monte Carlo simulation** involves repeated random sampling to explore the behavior of a system.

9.1.2. Generating Pseudorandom Numbers

To perform Monte Carlo simulations requires that computer programs have access to random numbers. The standard solution to this need is to actually use pseudorandom numbers, which are deterministically created lists of numbers that will pass a variety of statistical tests for lists of actual random numbers.

The pseudorandom number generators used by early computer programming languages used an algorithm known as a linear congruential generator. This algorithm is based on the following equation(Press et al., 1986).

$$I_j = (aI_{j-1} + c) \bmod m \qquad (9.1)$$

The constants a, c, and m are chosen to yield acceptable statistical properties for the sequence $I_0, I_1, I_2, \cdots$ To get the sequence started, the user usually provides a seed number, I_0.

There are recognized statistical problems with linear congruential generators if long sequences or subsets of sequences are required (Entacher, 1998). Modern numerical packages now use more sophisticated algorithms to produce pseudorandom number sequences including the Mersenne Twistor algorithm (Matsumoto & Nishimura, 1998) and the permuted congruential generator algorithm (O'Neill, 2014). To understand these more sophisticated algorithms requires the reader have more background in statistics and probability theory than this book assumes, so we will not discuss how they work.

The Python random module uses the Mersenne Twistor algorithm to generate pseudorandom numbers (Python Software Foundation, 2022). The numpy.random package uses the permuted congruential generator algorithm (NumPy Developers, 2022).

The following are the important functions from the random module that we use.

```
random.seed(a=None)
```

Initializes the random number generator. Any int can be used. If none is provided, then the system time is used. Using the same seed will generate the same sequence of pseudorandom numbers.

```
random.randint(a, b)
```

Returns a random integer N such that $a \leq N \leq b$.

```
random.random()
```

Returns the next random floating point number in the range [0.0, 1.0).

```
random.gauss(mu, sigma)
```

Returns a random number selected from the normal distribution, also called the Gaussian distribution. mu is the mean, and sigma is the standard deviation.

The following code is generally used to set up the random module for use in a program.

```
import random as rn
# initialize the generator
rn.seed(524287)
```

It is a good practice to explicitly set the seed so that a program gives reproducible results. This can be very helpful during the debugging phase of code development.

Numpy has its own random submodule. In versions before 1.17, the use of the Numpy random functions proceeded in much the same way as with the Python random module.

Numpy random module functions before Numpy version 1.17:

```
numpy.random.seed(a=None)
```

Initializes the random number generator. Any int can be used for the value of a. If none is provided, then the system time is used. Using the same seed will generate the same sequence of pseudorandom numbers.

```
numpy.random.randint(a, b, size=None)
```

Returns a random integer N such that $a \leq N \leq b$. The size keyword argument can be provided a tuple that specifies the shape of a numpy array of random integers that will be returned.

```
numpy.random.random(size=None)
```

Returns the next random floating point number in the range [0.0, 1.0). The size keyword argument can be provided a tuple that specifies the shape of a numpy array of random floats that will be returned.

```
numpy.random.normal(loc=0.0, scale=1.0, size=None)
```

Returns a random number selected from the normal distribution, also called the Gaussian distribution. loc is the mean, and scale is the standard deviation. The size keyword argument can be provided a tuple that specifies the shape of a numpy array of random floats selected from the normal distribution that will be returned.

Starting with Numpy version 1.17, while the functions listed above can still be used, the best use recommendation has changed. The goal of the new version usage is to avoid problems with the seed being unintentionally reset by complex codes. The new approach allows separate copies (instances, to use Object Oriented Programming language) of the random number generator to be created that are independent of random number generators used in other modules. The following code shows how to create a copy of the random number generator.

```
rng = numpy.random.default\_rng(seed=None)
```

The variable name rng will now refer to a separate copy of the numpy random module with scope restricted to the program unit in which it was defined. You can use any valid Python variable name instead of rng. If a seed number is not provided, then one generated from the computer operating system will be used.

Here is an example of code in a Jupyter notebook that uses the new Numpy random number generator approach.

```
[1]  1 import numpy as np
     2 # set up the numpy random number generator module
     3 np_rng = np.random.default_rng(524287)
     4
     5 # generate a random float
     6 random_float = np_rng.random()
     7 print(random_float)

0.8865893219051302
```

Figure 9.2.: Code segment in a Jupyter notebook for using the Numpy random number generator.

9.2. Random Walks

9.2.1. Definition

A **random walk** is a process in which an object moves away from a starting position at random. Lattice random walks are a simple case of random walks where the object is assumed to move on a regular grid of sites. It is a series of steps taken in random directions. The steps can be considered to be taken in any kind of space that is of interest. In the simplest case, the walk starts at zero and then at each step a +1 or -1 is added to the current position to yield an updated position. The walk would be the list of integers giving the position at each time.

More generally, a random walk on a lattice envisions the walk occurring on a regular grid such as the one-dimensional grid shown in Figure 9.2. In a lattice random walk, each direction for taking a step is assigned a probability and at each time step the walker chooses a direction according to the assigned probability. Figure 9.3 shows one possible series of steps.

A Monte Carlo simulation of a random walk will execute many separate walks and then perform statistical analyses of whatever property that is of interest.

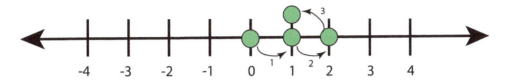

Figure 9.3.: Random walk on a one-dimensional lattice.

9.2.2. One-dimensional Random Walk

Let's consider a simple situation that would not be described well by a deterministic model. Consider a bar with its entrance in a narrow alley. If someone exits the bar, then they can only move right, left, or stay in place. Now, imagine a man has been consuming alcoholic beverages in the bar for a couple of hours. Needless to say, he is rather intoxicated, so he decides to walk home (at least he knows not to drink and drive!). He leaves the bar, which means stepping into the alley where he can only move left or right. Given his level of intoxication, the man cannot think clearly or even walk in a straight line. He steps erratically left or right with no discernable pattern to his choice. His walk in the alley is random.

We will not be able to model this man's walking behavior using a deterministic model, but we can still simulate the man's motion using a stochastic model and look for possible patterns in his behavior. Models that simulate the behavior of a system based on an element of randomness are called Monte Carlo simulations. We will create a Monte Carlo simulation of the drunk man's walking behavior called, appropriately enough, the random walk simulation. This type of model can be applied much more generally than simply describing the behavior of an intoxicated man, but this scenario is an easy one to visualize as we develop our understanding of stochastic simulations.

Recall our general problem-solving strategy:

Analysis: Analyze the problem

Design: Describe the data and develop algorithms

Implementation: Represent data using programming language data structures

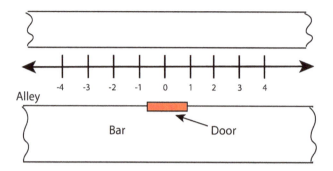

Figure 9.4.: Narrow alley for the drunk's walk. The origin of the axis is at the door to the bar.

and implement the algorithms with specific programming language code

Testing: Test and debug the code using test data

Let's analyze our problem: simulating the random walk of our intoxicated man. We note that the man can only move along the axis, so it is an inherently one-dimensional problem. The man will choose to take a step leftwards or rightwards with equal probability.

Now we will move on to designing our model. We will assume the man moves on the x-axis, as shown in Figure 9.4. His position at a particular time t is given by his x-coordinate, $x(t)$. We will make the following simplifying assumptions:

- The man attempts a step at equal time intervals.
- The man chooses to step left or right with equal probability, like a coin toss.
- All steps have length of 1 spatial unit.

Data structures:

N – total number of steps to simulate

i – the particular step currently being considered

xi – the position at step i

x_list – a list containing the position at each step

Here's an initial pseudocode solution to the simulation:

```
PROGRAM randomWalk1d
    Initialize the drunk's position at the origin: xi = 0
    Set up a list to contain the drunk's position at each step:
    defines x_list
    # set up a loop to execute N steps
    FOR i in the list (0,1,2,...N) DO
        Generate a random step
        Update xi
        Add xi to x_list
    ENDFOR
    Create a graph of position as a function of time
ENDPROGRAM
```

Let us take a more detailed look at the program step

Generate a random step

We will implement this step as a function named step(xi) that takes the current position xi and produces the next position, which is returned.

Here is a pseudocode version of the step(xi) function.

```
FUNCTION step
    INPUT: xi
    Choose a direction (left or right) at random
    IF left THEN
        xi = xi - 1
    ELSE
        xi = xi + 1
    ENDIF
    OUTPUT: xi
ENDFUNCTION
```

We will move on to the Implementation portion of our problem-solving strategy. Let us turn our attention to how we can generate random numbers in Python. The reality is that we cannot generate truly random numbers, but we can generate pseudorandom numbers that can pass tests for randomness as long as we do not try to generate too many of them.

We will use the Python random module to generate our pseudorandom numbers. We start our use of these functions with an import command:

import random as rn

One of the first steps in the main part of our program will be to set up the random number generator by providing a seed. We do this so that we can generate the same sequence of random numbers to aid in debugging the program.

So, how do we choose the left or right direction at random? We can use the random() function to generate a number between 0 and 1. If the number is less than 0.5 then we choose left for the direction. If the number is greater than or equal to 0.5 then we choose right for the direction. Here's Python code for the step(xi) function:

```python
def step(xi):
    r = rn.random()
    if r < 0.5:
        xi = xi -1
    else:
        xi = xi+1
    return xi
```

This code assumes the random module was imported and renamed rn by the main program.

A computer simulation of a random walk must assume a finite lattice size since the available memory for the simulation is finite. This fact means that we must decide how to handle the walker hitting a boundary. In the code being developed here, the step function must deal with the case of xi being on the boundary. There are two approaches to handling this situation. One is to assume reflective boundary conditions, which means that if the walker hits the boundary he just stays there. The other method is to use cyclic boundary conditions, which means that if the walker hits the boundary he actually goes to the other side, as if the x-axis wraps around on itself.

Let's implement the reflective boundary conditions. We need to define two more variables:

lx – the lowest x value

ux – the highest x value

Here is the revised step function:

```
def step(xi,lx,ux):
    r = rn.random()
    if r < 0.5:
        if xi >lx:
            xi = xi - 1
    else:
        if xi < ux:
            xi = xi + 1
    return xi
```

A complete code for performing this one-dimensional simulation is shown in Figure 9.5 The code is in the chapter file randomWalk1d.py. This code produces a plot of the position as a function of time (step number) shown in Figure 9.6. The mean and standard deviation of the position is also calculated in lines 41-43.

Another important random walk property to study is how the distance of the walker to the origin varies with time. For a one-dimensional random walk that starts at the origin, the distance to the origin is given by

$$D = |x| \tag{9.2}$$

Figure 9.7 shows the distance as a function of time for a one-dimensional random walk on a grid defined by $-1000 < x < 1000$.

While the distance appears to increase with time, there is a significant amount

```
1  """
2  Program: 1-d Random Walk
3  Author: C.D. Wentworth
4  Version: 9.23.2022.1
5  Summary:
6          This program performs a 1-dimensional random
7          walk simulation.
8  History:
9      9.23.2022.1: base
10 """
11 import matplotlib.pyplot as plt
12 import numpy as np
13 import random as rn
14
15 def step(xi,lx,ux):
16     import random as rn
17     r = rn.random()
18     if r < 0.5:
19         if xi > lx:
20             xi = xi -1
21     else:
22         if xi < ux:
23             xi = xi+1
24     return xi
```

Figure 9.5a: Python code for the 1D random walk.

```python
25  # Main Program
26
27  # set up the grid
28  lx = -100
29  ux = 100
30
31  # set up random generator
32  rn.seed(524287)
33  N = 200
34
35  # execute random walk
36  xi = 0
37  xlist = [xi]
38  for i in range(N):
39      xi=step(xi,lx,ux)
40      xlist.append(xi)
41  xnp = np.array(xlist)
42  xmean = xnp.mean()
43  xsd = np.sqrt(xnp.var()/float(N-1))
44
45  # create plot of x - t
46  plt.plot(xlist)
47  plt.xlabel('t [steps]', fontsize=14)
48  plt.ylabel('x', fontsize=14)
49  plt.title('Position - Time', fontsize=18)
50  plt.savefig('randomwalk1d.png', dpi=300)
51  plt.show()
52
53  # print position
54  print('mean x = ',xmean,' sd = ',xsd)
55  print('final position x = ', xlist[-1])
```

Figure 9.5b: Continuation of Python code for the 1D random walk.

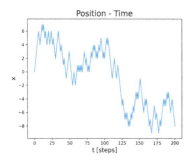

Figure 9.6.: Plot of position versus time for one simulation.

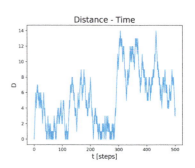

Figure 9.7.: Distance-time for a one-dimensional random walk.

Table 9.1.: Distance array. N = number of steps in a simulation. S = total number of simulations.

D11	D12	D13	...	D1S
D21	D22	D23	...	D2S
D31	D32	D33	...	D3S
.	.	.	.	.
.	.	.	.	.
.	.	.	.	.
DN1	DN2	DN3	...	DNS

of noise in this property. This is a typical result for any measured property in a stochastic simulation just as it is for a measurement taken in a real world experiment. To obtain a better picture of the distance as a function of time we should repeat the experiment, that is the simulation, many times and calculate the mean distance as a function of time where the mean is calculated from our sample of many simulations.

The key data structure for calculating the mean distance as a function of time is a numpy array that contains the distance to the origin at each time for each of the separate simulations. Table 9.1 shows the structure of this array.

$$D_{tn} = \text{distance at time } t \text{ for simulation } n \tag{9.3}$$

The mean distance at any particular time is calculated by averaging over all the columns of the row corresponding to the time of interest. Figure 9.8 shows the revised code that performs repetitions of the simulation and then calculates the mean distance as a function of time. Figure 9.8 shows a plot of the mean distance as a function of time (number of steps). Most of the random variation shown in Figure 9.6 has disappeared.

```
1  """
2  Program: 1-d Random Walk with Distance Calculation
3  Author: C.D. Wentworth
4  Version: 9.24.2022.1
5  Summary:
6          This program performs a 1-dimensional random
7          walk simulation. It also calculates the distance
8          from the origin at each time step.
9  History:
10     9.24.2022.1: base
11  """
12 import matplotlib.pyplot as plt
13 import numpy as np
14 import random as rn
15
16 def step(xi,lx,ux):
17     import random as rn
18     r = rn.random()
19     if r < 0.5:
20         if xi > lx:
21             xi = xi -1
22     else:
23         if xi < ux:
24             xi = xi+1
25     return xi
```

Figure 9.8a: Multi-simulation version of the one-dimensional random walk.

```
29  # Main Program
30
31  # set up the grid
32  lx = -1000
33  ux = 1000
34
35  # set up random generator
36  rn.seed(524287)
37  N = 500
38
39  # execute random walk
40  xi = 0
41  xlist = [xi]
42  D_list = [0]
43  for i in range(N):
44      xi=step(xi,lx,ux)
45      xlist.append(xi)
46      D = np.abs(xi)
47      D_list.append(D)
48  xnp = np.array(xlist)
49  xmean = xnp.mean()
50  xsd = np.sqrt(xnp.var()/float(N-1))
51
52  # create plot of D - t
53  plt.plot(D_list)
54  plt.xlabel('t [steps]', fontsize=14)
55  plt.ylabel('D', fontsize=14)
56  plt.title('Distance - Time', fontsize=18)
57  plt.savefig('randomwalk1d_D-t.png', dpi=300)
58  plt.show()
59
60  # print position
61  print('mean x = ',xmean,' sd = ',xsd)
62  print('final position x = ', xlist[-1])
```

Figure 9.8b: Continuation of Multi-simulation version of the one-dimensional random walk.

9.2.3. Two-dimensional Random Walk

Let's give our intoxicated walker more room to roam around. We will place him in a large parking lot so that he can move about on a plane, a two-dimensional surface. We want to simulate his trajectory on the parking lot. Our model will be based on some simplifying assumptions, similar to the one-dimensional walk.

■ The man attempts a step at equal time intervals.

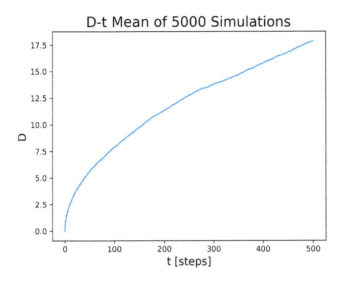

Figure 9.9.: Mean distance as a function of time from 5000 simulation repetitions.

■ The man chooses to step north, south, east, or west with equal probability.

■ All steps have length of 1 spatial unit.

With these assumptions, our walker will roam about on a grid like chess pieces on a game board. The basic algorithm for performing the simulation will be the same as for the one-dimensional case. The major difference will be in the step function. It must now choose one of the four directions at random instead of either right or left. Here is a pseudocode version of the step function.

```
FUNCTION step
    INPUT: xi,yi
    Pick a random number r, from (1,2,3,4)
    IF r is 1 THEN
        Step east by incrementing xi
    ELIF r is 2 THEN
        Step west by decreasing xi
    ELIF r is 3 THEN
        Step north by increasing yi
    ELIF r is 4 THEN
        Step south by decreasing yi
    ENDIF
    OUTPUT: xi,yi
ENDFUNCTION
```

We must implement boundary conditions since the simulated grid will be finite. This will require minimum and maximum values for both the x and y coordinates. These values will be defined using the variables lx,ux,ly,uy. Figure 9.10 gives the Python code for the two-dimensional step function with reflective boundaries. Figure 9.10 shows an example of the overall trajectory of this random walk.

```python
"""
Title: Random Walk in 2D: base
Author: C.D. Wentworth
version: 3-21-2019.1
Summary: This program performs a random walk on a
         two-dimensional lattice. It uses reflective
         boundary conditions.
version history:
         3-21-2019.1: base
"""
import matplotlib.pylab as plt
import numpy as np
import random as rn
import turtle as trt


def step(xi, yi, lx, ux, ly, uy):
    import random as rn
    r = rn.randint(1, 4)
    if r == 1:
        # go east
        if xi < ux:
            xi = xi + 1
    elif r == 2:
        # go west
        if xi > lx:
            xi = xi - 1
    elif r == 3:
        # go north
        if yi < uy:
            yi = yi + 1
    else:
        # go south
        if yi > ly:
            yi = yi - 1
    return xi, yi
```

Figure 9.10a: Code for random walk on a 2-d lattice.

```
39  # --Main Program
40  # set up random generator
41  rn.seed(42)
42
43  # define the grid
44  lx = -40
45  ux = 40
46  ly = -40
47  uy = 40
48
49  # set up the simulation
50  N = 1000
51  xi = 0
52  yi = 0
53  position_array = np.zeros((N + 1, 2))
54
55  # execute random walk
56  for i in range(1, N + 1):
57      xi, yi = step(xi, yi, lx, ux, ly, uy)
58      position_array[i, 0] = xi
59      position_array[i, 1] = yi
60
61  x_array = position_array[:, 0]
62  y_array = position_array[:, 1]
63
64  # create plot
65  x0 = x_array[0]   # initial x
66  y0 = y_array[0]   # initial y
67  xf = x_array[-1]  # final x
68  yf = y_array[-1]  # final y
69  plt.plot(x_array, y_array)
70  plt.plot(x0, y0, linestyle='', marker='o', markersize=8,
71          markeredgecolor='green', markerfacecolor='green')
72  plt.plot(xf, yf, linestyle='', marker='o', markersize=8,
73          markeredgecolor='red', markerfacecolor='red')
74  plt.xlim(-20, 20)
75  plt.ylim(-20, 20)
76  plt.savefig('randomWalk2d.png', dpi=300)
77  plt.show()
```

Figure 9.10b: Continuation of Code for random walk on a 2-d lattice.

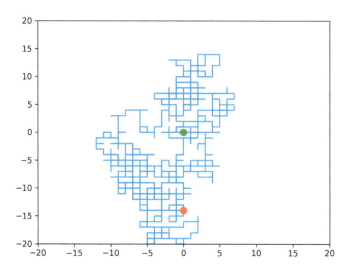

Figure 9.11.: Trajectory of a random walk on a 2-d lattice. Green shows origin. Red shows final position.

9.3. Cellular Automaton Model

Cellular automata can be considered as a particular abstract framework for describing the real world, that is, a modelling framework, or as a kind of mathematical object that can be used to build models or investigate computational problems (Berto & Tagliabue, 2022). We will start with a formal, mathematical definition of a cellular automaton. A cellular automaton model is defined as

- a discrete lattice of sites or cells in d dimensions, L

- operating using discrete time steps

- a discrete set of possible cell values, S

- a defined neighborhood, N of each cell in L

- a transition rule, Φ, that specifies how the state of a cell will be updated depending on its current value and the values of sites in the neighborhood.

We will consider the cellular automaton to be the collection L, S, N, Φ. A specific example will be defined in the next section, which should make this definition less abstract. Cellular automata have been used to model

- computational machines
- growth of biological systems such as biofilms
- evolution of life
- phase transitions in magnetic materials
- urban development

Cellular automata are often used to show how complex patterns can be the result of simple rules governing the behavior of the system being considered.

9.3.1. Deterministic Cellular Automata

We will define a simple one-dimensional cellular automaton. The lattice L is an infinite, one-dimensional set of cells, as shown in Figure 9.12.

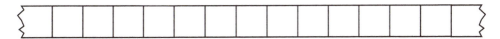

Figure 9.12.: One-dimensional lattice of cells for a cellular automaton.

The possible cell values will be $S = \{0, 1\}$. We will visualize a cell with state 0 as white and state 1 as black. The neighborhood, N, of a cell will be defined as the nearest neighbors of the cell: the cell to the left and the cell to the right. A particular combination of neighbors and center will be symbolized by LCR.

The transition rule will be a new assignment of the C state based on the current value of the LCR combination. There are eight possible LCR combinations for this class of cellular automata, shown in Table 9.2. There are two possible new values for the C cell given a specific LCR combination. Therefore, there are 2^8 possible transition rules for this class of cellular automata.

We will enumerate each possibility by an integer from 1 to 256. The transition

Table 9.2.: A specific transition rule (Rule 150) for the class of cellular automata being considered.

L	C	R	new C
1	1	1	1
1	1	0	0
1	0	1	0
1	0	0	1
0	1	1	0
0	1	0	1
0	0	1	1
0	0	0	0

rule will be given by the binary representation of rule number given by the base 10 integer. For example, Rule 150 has the binary representation 10010110. Each of the eight digits in this representation gives the transition rule a specific LCR combination. Table 9.2 illustrates how this assignment can be made.

Now we must construct a computer simulation of this simple cellular automaton model. We will base the code on one developed by Cyrille Rossant (Rossant, 2018).

Since the cellular automaton is being simulated by a computer, the lattice used to define the automaton must by finite. This means we must specify how the neighbors are defined for the cells all the way on the left and all the way on the right of the lattice. We will use periodic boundary conditions. This means that for the cell all the way on the left, the left neighbor will be the cell all the way on the right of the lattice. For the cell all the way on the right, the right neighbor will be the cell all the way on the left of the lattice. Figure 9.13 shows the code contained in the file CA_sim.py for simulating the simple one-dimensional cellular automaton. Figure 9.14 shows the output from the simulation for a Rule 150 cellular automaton with an initial state comprised of one active cell in the middle of the lattice.

Table 9.3.: Data structures required for the one-dimensional cellular automaton simulation.

Variable	Type	Description
size	int	The number of cells in the lattice.
steps	int	The number of time steps to simulate.
rule	int	The state transition rule given as a decimal number.
rule_b	ndarray	The state transition rule given as a binary number represented in a numpy array.
x	ndarray	A 2d numpy array where each row gives the state of the cellular automaton at a specific time. The array contains the entire simulated evolution of the automaton.
LCR_array	ndarray	Each column of the array gives the Left, Center, and Right state values for a particular cell in the automaton. Used in step function.
xs	ndarray	A 1-dim array that gives the cell states for each cell in the automaton at one time. Used in the step function.

```python
"""
Program: 1-d Cellular Automaton Simulation
Author: Based on code from Rossant, C. (2018).
        IPython Interactive Computing and Visualization Cookbook-
        Second Edition. Packt Publishing.
        Revisions by C.D. Wentworth
Version: 9.29.2022.1
Summary:
        This program simulates a one-dimensional deterministic
        cellular automaton.
History:
    9.29.2022.1: base
"""
import numpy as np
import matplotlib.pyplot as plt

u = np.array([[4], [2], [1]])

def step(xs, rule_b):
    """
    Compute a single state of an elementary cellular
    automaton.
    """
    # Define the array that gives the LCR state values
    # for each cell.
    LCR_array = np.vstack((np.roll(xs, 1), xs,
                          np.roll(xs, -1))).astype(np.int8)
    # We get the LCR pattern numbers between 0 and 7
    # for each cell in the automaton.
    z = np.sum(LCR_array * u, axis=0).astype(np.int8)
    # We get the updated cell states given by the rule.
    return rule_b[7 - z]
```

Figure 9.13a: One-dimensional cellular automaton simulation code.

```
34  def generate(rule, size=100, steps=100):
35      """
36      Simulate an elementary cellular automaton given
37      its rule (number between 0 and 255).
38      """
39      # Compute the binary representation of the rule.
40      rule_b = np.array(
41          [int(_) for _ in np.binary_repr(rule, 8)],
42          dtype=np.int8)
43      x = np.zeros((steps, size), dtype=np.int8)
44      # Define the initial state
45      x[0, 50] = 1
46      # Apply the step function iteratively.
47      for i in range(steps - 1):
48          x[i + 1, :] = step(x[i, :], rule_b)
49      return x
50
51
52  # Main Program
53  rule = 30
54  size = 100
55  steps = 50
56  x = generate(rule, size, steps)
57  plt.matshow(x, cmap=plt.cm.binary)
58  title_string = 'Rule ' + str(rule)
59  plt.title(title_string, fontsize=18)
60  fig_file = 'CA_rule' + str(rule) + '.png'
61  plt.savefig(fig_file, dpi=300)
62  plt.show()
```

Figure 9.13b: Continuation of One-dimensional cellular automaton simulation code.

9.3.2. Probabilistic Cellular Automata

The cellular automata considered in the previous section are completely deterministic. Once the transition rule and initial state have been chosen the evolution of the lattice is determined. Repetition of the evolution will always yield the same result. We can introduce randomness in at least two ways:

1. create a random initial state or
2. establish a transition rule that updates the state of a cell using the neighborhood state and then a new cell state chosen according to a probability distribution.

Probabilistic cellular automata are usually defined using the second method of

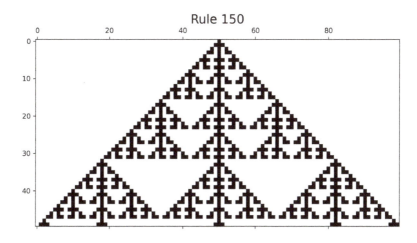

Figure 9.14.: This figure shows the 1-d Rule 150 automaton simulated for 50 time steps.

achieving randomness. We will not discuss coding of a probabilistic cellular automaton simulation here.

9.4. Computational Problem Solving: Comparing Betting Strategies at the Roulette Wheel

We began this chapter with the problem of comparing betting strategies when playing roulette. Let us return to that problem and develop a Monte Carlo simulation approach to solving the problem. We will consider playing American roulette and will restrict the betting types to a small subset of the possible ones.

9.4.1. Analysis

American roulette uses a wheel with 38 pockets that include both a 0 and a 00 pocket. The 0 and 00 pockets are green. The other pockets alternate between black and red. There are a large number of possible bets, so we will need to restrict the analysis to a small subset to facilitate the initial solution ("Roulette," 2022). For this initial analysis, we will consider the following types of bets.

1. Straight/Single: bet on a single number; 35:1 payout
2. Red/Black: bet on either red or black; 1:1 payout
3. Green: bet on green; 17:1 payout

Each pocket of the wheel has numerical value, including 00, and a color. Table 9.4 gives the sequence of the slots going clockwise around the wheel. We will assume that when the wheel spins, the ball has an equal chance of falling into any one of the 38 pockets. For this initial attempt at simulating roulette, we will assume that each play involves a constant bet and a choice betting type. A game will be defined as a series of plays. For a game, the series of plays will end when the players balance goes below zero or if the balance goes above a specified balance value. Our gambler will continue playing until either they run out of money or until they achieve a specified level of profit.

Table 9.4.: Layout of the American Roulette wheel, going clockwise.

pocket index	value	color	pocket index	value	color
0	0	green	19	0	green
1	28	black	20	27	red
2	9	red	21	10	black
3	26	black	22	25	red
4	30	red	23	39	black
5	11	black	24	12	red
6	7	red	25	8	black
7	20	black	26	19	red
8	32	red	27	31	black
9	17	black	28	18	red
10	5	red	29	6	black
11	22	black	30	21	red
12	34	red	31	33	black
13	15	black	32	16	red
14	3	red	33	4	black
15	24	black	34	23	red
16	36	red	35	35	black
17	13	black	36	14	red
18	1	red	37	2	black

9.4.2. Design

Let us start the design of the algorithm by defining data structures that will be required. We will also create the roulette wheel data from Table 9.4 as columns in a tab-delimited text file. This file will be named wheel.txt. Table 9.5 gives a list of data structures that will be required for the roulette simulation.

Table 9.5.: Data structures for the roulette simulation.

Variable	Type	Description
balance	int	This contains the current balance value of the gambler's purse.
stop_balance	int	The balance value that will stop the game. Assumed to be larger than the starting value
bet_amount	int	
bet_type	str	Specifies whether the gambler is betting on a color ('black', 'red', green') or a specific value ('0', '1', ..., '00')
number_of_games	int	The number of repeated games to play, which will create the Monte Carlo simulation sample.
bet_types	dictionary	Associates the payout value with a bet type.
df	pandas data frame	Contains the wheel information read in from the wheel.txt file.
values	list	Contains all of the pocket values specified in the order found on the wheel going clockwise.
colors	list	Contains all of the pocket colors specified in the order found on the wheel going clockwise.
balance_series	list	Contains the balance value for each play in a game
spin_value	str	Contains the pocket value after a spin of the wheel.
spin_color	str	Contains the pocket color after a spin of the wheel.

The following gives a pseudocode version of the main program for the simulation.

```
PROGRAM American_Roulette_Simulation-ConstantBetting
    Specify gambler's data (balance, stop_balance, bet_amount,
                            bet_type, number_of_games)
    Initialize the random number generator
    Define the bet_types dictionary
    Define the wheel variables (values, colors)
    FOR g IN list_of_games DO
        Play a game
        Create a plot of the balance versus number of plays
    ENDFOR
ENDPROGRAM
```

The Play a game line will require another function, which we name game_constant_bet, and this function will require another function called make_a_play.

```
FUNCTION game_constant_bet
    INPUT: balance, stop_balance, bet_amount, bet_type, bet_types,
           values, colors
    balance_series = [balance]
    WHILE (balance > 0) and (balance < stop_balance) DO
        balance = make_a_play(balance, bet_amount, bet_type,
                              bet_types, values, colors)
        balance_series.append(balance)
    ENDWHILE
    OUTPUT: balance_series
ENDFUNCTION
```

The function game_constant_bet requires another function make_a _play.

```
FUNCTION make_a_play
    INPUT: balance, bet_amount, bet_type, bet_types, values, colors
    Perform a spin of the wheel (defines spin_value, spin_color)
    IF bet_type == 'black'
        IF spin_color == 'black'
            balance = balance + bet_amount*bet_types['black']
        ELSE
            balance = balance - bet_amount
        ENDIF
    ELIF bet_type == 'red'
        IF spin_color == 'red'
            balance = balance + bet_amount*bet_types['red']
        ELSE
            balance = balance - bet_amount
        ENDIF
    ELIF bet_type == 'green'
        IF spin_color == 'green'
            balance = balance + bet_amount*bet_types['green']
        ELSE
            balance = balance - bet_amount
```

```
        ENDIF
    ELSE
        IF spin_value == bet_type
            balance = balance + bet_amount*bet_types['value']
        ELSE
            balance = balance - bet_amount
        ENDIF
    ENDIF
    OUTPUT: balance
ENDFUNCTION
```

The function make_a_play requires a function which we call spin.

```
FUNCTION spin
    INPUT: values, colors
    Get a random integer between 0 and 37 (defines i)
    value = values(i)
    color = colors(i)
    OUTPUT: value, color
ENDFUNCTION
```

These pseudocode versions of the algorithm should allow an easy translation into Python code.

9.4.3. Implementation

Figure 9.15 gives the Python implementation of the American Roulette with Constant Bet simulation.

```python
"""
Program: American Roulette with Constant Bet
Author: C.D. Wentworth
Version: 9.1.2022.1
Summary:
        This program performs a Monte Carlo simulation of the
        American roulette game using a constant bet strategy based
        either color choice or specific value choice.
History:
    9.1.2022.1: base
"""
import matplotlib.pyplot as plt
import numpy as np
import pandas as pd

def spin(values,colors):
#     i = rn.randint(0, 37)
    i = rng.integers(low=0 , high=37 )
    value = values[i]
    color = colors[i]
    return value, color

def make_a_play(balance, bet_amount, bet_type, bet_types,
                values, colors):
    spin_value, spin_color = spin(values, colors)
    if bet_type == 'black':
        if spin_color == 'black':
            balance = balance + bet_amount*bet_types['black']
        else:
            balance = balance - bet_amount
    elif bet_type == 'red':
        if spin_color == 'red':
            balance = balance + bet_amount*bet_types['red']
        else:
            balance = balance - bet_amount
    elif bet_type == 'green':
        if spin_color == 'green':
            balance = balance + bet_amount*bet_types['green']
        else:
            balance = balance - bet_amount
    else:
        if spin_value == bet_type:
            balance = balance + bet_amount*bet_types['value']
        else:
            balance = balance - bet_amount
    return balance
```

Figure 9.15a: Python implementation of the American Roulette with Constant Bet
simulation.

257

```python
48  def game_constant_bet(balance, stop_balance, bet_amount, bet_type,
49                        bet_types, values, colors):
50      balance_series = [balance]
51      while (balance > 0) and (balance < stop_balance):
52          balance = make_a_play(balance, bet_amount, bet_type, bet_types,
53                      values, colors)
54          balance_series.append(balance)
55      return balance_series
56
57  # Main Program
58  # Initialize gambler's data
59  balance = 1000
60  stop_balance = 1200
61  bet_amount = 10
62  bet_type = 'green'
63  number_of_games = 100
64
65  # Set up the numpy random number generator
66  rng = np.random.default_rng(314159)
67
68  # Initialize the bet type payout
69  bet_types = {'black': 1, 'red': 1, 'green': 17, 'value': 35  }
70
71  # Read in the wheel data
72  df = pd.read_csv('wheel.txt', sep='\t', header=1,
73          dtype={'index':int, 'value':'string', 'color':'string'})
74
75  # Convert pandas columns to tuples
76  values = tuple(df['value'].tolist())
77  colors = tuple(df['color'].tolist())
78
79  for g in range(number_of_games):
80      balance_series = game_constant_bet(balance, stop_balance,
81                                  bet_amount, bet_type,
82                                  bet_types, values, colors)
83      plt.plot(balance_series)
84  plt.xlabel('play')
85  plt.ylabel('balance')
86  title_string = 'Constant Bet Strategy: ' + bet_type
87  plt.title(title_string, fontsize=18)
88  plt.savefig('roulette_constantBet_value15_1200.png', dpi=300)
89  plt.show()
```

Figure 9.15b: Continuation of Python implementation of the American Roulette with Constant Bet simulation.

This code is in the file roulette_constantBet.py.

Table 9.6.: Generated test data from the spin function. A comparison is made with Table 9.4 to determine correctness.

pocket index	pocket value	pocket color	correct entry
29	6	black	yes
4	30	red	yes
14	3	red	yes
32	16	red	yes
31	33	black	yes
28	18	red	yes
29	6	black	yes
2	9	red	yes
27	31	black	yes
3	26	black	yes

9.4.4. Testing

A good strategy for testing the program is to create a series of small programs that test the individual functions. If the individual functions work as expected, then you will have some significant evidence that the overall program works as expected. Figure 9.16 shows a short code that will test the spin function. A write statement is inserted into the function to write out some test data to a file. Table 9.6 shows data generated by this code. The rows are compared with Table 9.4. After the test, lines 10-13 are deleted.

Similar short programs can be used to test the other functions.

To obtain some initial data for the constant bet strategy we will choose the following initial gambler's data:

balance = 1000
stop_balance = 1200
bet_amount = 10
number_of_games = 100

We will run the simulation for the following bet types: 'red', 'green', '15', and com-

```
1  """
2  Program: Test Spin Function
3  Author: C.D. Wentworth
4  Version: 9.1.2022.1
5  Summary:
6          This program tests the spin function from the American
7          Roulette Constant Bet simulation.
8  History:
9      9.1.2022.1: base
10 """
11 import matplotlib.pyplot as plt
12 import numpy as np
13 import pandas as pd
14 import random as rn
15
16 def spin(values,colors):
17     i = rn.randint(0, 37)
18     value = values[i]
19     color = colors[i]
20     # write test data
21     write_string = str(i) + '\t' + value + '\t' + color + '\n'
22     out_file.write(write_string)
23     # end write test data
24     return value, color
25
26 # Main Program
27 number_of_tests = 10
28
29 # Initialize random number generator
30 rn.seed(524287)
31
32 # Read in the wheel data
33 df = pd.read_csv('wheel.txt', sep='\t', header=1,
34         dtype={'index':int, 'value':'string', 'color':'string'})
35
36 # Convert pandas columns to tuples
37 values = tuple(df['value'].tolist())
38 colors = tuple(df['color'].tolist())
39
40 out_file = open('testSpinFunctionData.txt', 'w')
41 for t in range(number_of_tests):
42     value, color = spin(values, colors)
43 out_file.close()
```

Figure 9.16.: Short program to test the spin function.

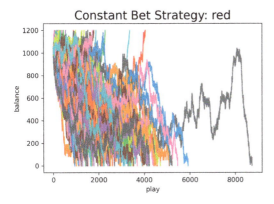

Figure 9.17.: Monte Carlo simulation of roulette with a constant bet on red strategy.

pare the plots of balance as a function of play. Figures 9.17 – 9.19 show the results of these simulations. Each simulation is for 100 games. Based on this data, a constant bet on red allows one to play longer before running out of money than the other bet types. A more thorough analysis that explores bet size effects and calculates the actual probability of reaching 1200 should be done before recommending the best version of the constant bet strategy.

Figure 9.18.: Monte Carlo simulation of roulette with a constant bet on green strategy.

d

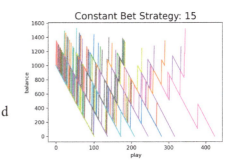

Figure 9.19.: Monte Carlo simulation of roulette with a constant bet on the value 15 strategy.

9.5. Program Modification Problems

1. Starting with the two-dimension random walk code in Ch9ProgModProb1.py, write a program that calculates the distance to the origin of the walker as a function of step number (time). The program should produce a plot of D versus t. You can remove the plot of the actual walk.

2. Starting with the one-dimensional cellular automata simulation code in Ch9ProgModProb2.py, create a program that assigns an initial state of 1 to several randomly selected cells. Explore the differences in the patterns obtained between the revised program and the original one with initial state of 1 in the middle of the lattice.

3. Starting with the American Roulette with Constant Bet code in Ch9ProgModProb3.py, create a program to calculate the proportion of games that end with a profit for the bet on green and bet on a specific value bet types.

9.6. Program Development Problems

1. Create a program that simulates a two-dimensional random walk repeatedly and calculates the mean distance at each step over the sample of repeated simulations and produces a plot of mean distance as a function of step. The sample size should be large enough so that you get a smooth curve.

2. Create a program that will simulate the American roulette game using the Martingale strategy that involves doubling the bet value after losing a play. Explore the behavior of this strategy for the different betting types considered in the program American Roulette with Constant Bet contained in the file roulette_constantBet.py.

9.7. References

Berto, F., & Tagliabue, J. (2022). Cellular Automata. In E. N. Zalta (Ed.), *The Stanford Encyclopedia of Philosophy* (Spring 2022). Metaphysics Research Lab, Stanford University. https://plato.stanford.edu/archives/spr2022/entries/cellular-automata/

Derek Lynn. (2020). *Roulette wheel in a casino for gambling entertainment* [Photograph]. https://unsplash.com/photos/mD1V-eS1Wb4

Entacher, K. (1998). Bad subsequences of well-known linear congruential pseudorandom number generators. *ACM Transactions on Modeling and Computer Simulation, 8*(1), 61–70. https://doi.org/10.1145/272991.273009

Matsumoto, M., & Nishimura, T. (1998). Mersenne twister: A 623-dimensionally equidistributed uniform pseudo-random number generator. *ACM Transactions on Modeling and Computer Simulation, 8*(1), 3–30. https://doi.org/10.1145/272991.272995

NumPy Developers. (2022). *Random Generator—NumPy v1.23 Manual.* https://numpy.org/doc/stable/reference/random/generator.html

O'Neill, M. E. (2014). *PCG: A Family of Simple Fast Space-Efficient Statistically Good Algorithms for Random Number Generation* (HMC-CS-2014-0905). Harvey Mudd College Computer Science Department. https://www.pcg-random.org/paper.html

Press, W. H., Teukolsky, S. A., Vetterling, W. T., & Flannery, B. P. (1986). *Numerical Recipes: The Art of Scientific Computing.* http://www.librarything.com/work/249712/book/225760462

Python Software Foundation. (2022). *Random—Generate pseudo-random numbers—Python 3.10.7 documentation.*

https://docs.python.org/3/library/random.html

Rossant, C. (2018). *IPython Interactive Computing and Visualization Cookbook—Second Edition.* Packt. https://www.packtpub.com/product/ipython-interactive-computing-and-visualization-cookbook-second-edition/9781785888632

Roulette. (2022). In *Wikipedia.* https://en.wikipedia.org/w/index.php?title=Roulette&oldid=1106799426

10. Fitting Models to Data: The Method of Least Squares

Motivating Problem: Damped Spring

Figure 10.1 shows the setup for a physics experiment involving a mass m (= 0.550 [kg]) attached to a spring. The spring is stretched from rest and released. Position data collected by a motion detector is shown in Figure 10.2. Since the data shows significant damping, the situation will require more than simply the ideal force law (Hooke's Law: $F = -ky$) to understand. By adding a drag force proportional to velocity, $-bv_y$, an appropriate model can be constructed. The solution to Newton's 2nd Law is discussed in most general physics textbooks and yields (Knight, 2016)

$$y(t) = Ae^{-bt/2m} \cos(\omega t + \phi_0)$$

$$\omega = \sqrt{\omega_0^2 - \frac{b^2}{4m^2}}, \omega_0 = \sqrt{\frac{k}{m}}$$

(10.1)

In Equation (10.1), $y(t)$ is the position measured with respect to the mass's equilibrium position.

Our computational problem will be to choose the best-fit values for the model parameters, A, b, and k, so that the model equation fits the provided data.

10.1. Introduction

In Chapter 7 on dynamical systems, we considered a mathematical model for the growth of bacteria. Our development of the model was guided by some actual data on growth for a couple of different species of bacteria. When we needed to choose values for parameters that appeared in the model, we used a trial-and-error

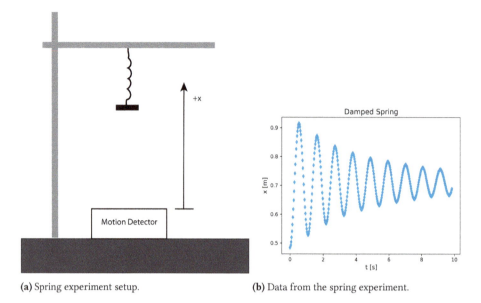

(a) Spring experiment setup.   **(b)** Data from the spring experiment.

Figure 10.1.: Damped spring experiment.

method to choose values that gave a good fit of the model to the available data. We need a more statistically sophisticated method that can choose best-fit values and estimate how much experimental uncertainty the values have.

10.2. The Method of Least Squares

10.2.1. Definition

While there are several methods that can be used to statistically estimate best-fit model parameter values, we will use the method of least squares. It is a widely used method and usually a good first-choice for the models we often encounter in science and engineering. We will introduce the basic framework with some formal definitions and then move on to implementing the method computationally, first with simple linear models and then with more complex non-linear models.

Let us assume that we have a model with independent variables $\mathbf{x} = (x_1, x_2, \cdots, x_v)$, a set of model parameters $\mathbf{p} = (p_1, p_2, \cdots, p_m)$, and the model function $f(\mathbf{x}; \mathbf{p})$ that determines the dependent variable y.

$$y = f(\mathbf{x}; \mathbf{p}) \tag{10.2}$$

In the simplest case there will be just one independent variable, but the method works for more complex cases.

Suppose we have performed an experiment that produced n data points

$$(\mathbf{x}_1, y_1), \cdots, (\mathbf{x}_n, y_n)$$

Next, we define a function called the Least Square Estimator, $L(\mathbf{p})$, by the equation

$$L(\mathbf{p}) = \sum_{i=1}^{n} (y_i - f(\mathbf{x}_i; \mathbf{p}))^2 \tag{10.3}$$

Note that L is a function of the model parameters. The least squares method for data fitting involves choosing the model parameter values $\mathbf{p}$ that minimize the function L.

10.2.2. Linear Regression

If we are using a simple linear model

$$y = ax + b \tag{10.4}$$

then statisticians have derived closed-form formulae for the best-fit slope, a, and intercept, b. They use the calculus principle that necessary conditions for a function to be a minimum are that first partial derivatives with respect to the independent variables must be zero. This is symbolized by

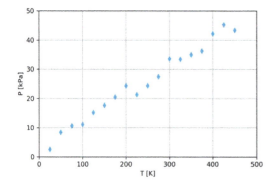

Figure 10.2.: Pressure data as a function of temperature for a fixed volume of gas.

$$\frac{\partial L(a,b)}{\partial a} = 0$$

$$\frac{\partial L(a,b)}{\partial b} = 0$$

(10.5)

These equations can be solved for the best-fit values of a and b. We will not work through the calculus or give the explicit formulae. You will learn to do that in a more advanced statistics course (Akritas, 2018, p. 6). Instead, we will learn how to implement the linear least squares technique numerically, letting the computer do all of the work.

We will use a specific example to learn this technique. Consider a rigid container with some air in it. We measure the pressure of this gas as the temperature is varied. The volume is held constant. Figure 10.2 shows some of the data from the file GasData.txt.

Visual inspection of the figure suggests that a linear model should fit this data rather well. We will use the method of least squares, linear regression version, to find the best fit slope and intercept. We expect the data presented in Figure 10.2 to fit the following model equation:

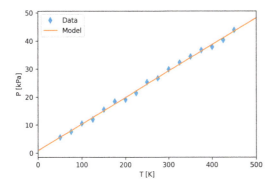

Figure 10.3.: Gas data with a linear model.

$$P = aT + b \qquad (10.6)$$

Minimizing the Least Squares Estimator function involves minimizing the vertical distance between the data points and the theoretical model value. Consider Figure 10.3, which shows our data with a linear model drawn as a red line.

The method of least-squares will find the model parameters, the slope and intercept in this case, that will minimize the sum of the squares of the residuals. The model line shown in Figure 10.3 was produced by the linear regression formulae from the least squares method. It is hard to imagine a different line that would come closer to the data points.

We will make use of a Python library function that will apply the linear regression formulae to the data set and give the user the optimized slope and intercept. The function is from the scipy.stats library. It is the scipy.stats.linregress() function (The SciPy Community, 2022b).

Syntax:

```
out = scipy.stats.linregress(x,y)
```

x = a list with the independent variable values

y = a list with corresponding dependent variable values

The function returns a list-like data structure, called out above, containing the following values (data type in parentheses)

slope : (float) slope of the regression line

intercept : (float) intercept of the regression line

r-value : (float) correlation coefficient

p-value : (float) two-sided p-value for a hypothesis test whose null hypothesis is that the slope is zero.

stderr : (float) Standard error of the slope estimate

The following lines of code will retrieve all of these values from the variable out.

```
slope = out[0]
intercept = out[1]
rvalue = out[2]
pvalue = out[3]
se = out[4]
```

Figure 10.4 shows a Python script that will apply the linregress function to the gas data. The code produces the following text output.

slope= 0.0948

intercept= 0.8803

rvalue= 0.9986

pvalue= 1.32e-20

se= 0.0013

Figure 10.3 shows the graph produced by the code.

```python
1  """
2  Program: Gas Data Fit
3  Author: C.D. Wentworth
4  Version: 3-20-2017.1
5  Summary:
6      This program reads in pressure-temperature data and fits
7      the data to a linear model using the scipy.stats.linregress()
8      function.
9  Version History:
10     3-20-2017.1: Base
11
12 """
13
14 import numpy as np
15 import matplotlib.pyplot as plt
16 import scipy.stats as ss
17 # import data
18 cols = np.loadtxt('GasData.txt',  skiprows=3)
19 tData = cols[:, 0]
20 PData = cols[:, 1]
21 # Fit data to linear model
22 out = ss.linregress(tData, PData)
23 slope = out[0]
24 intercept = out[1]
25 rvalue = out[2]
26 pvalue = out[3]
27 se = out[4]
28 print('slope= ', format(slope,'7.4f'))
29 print('intercept= ',format(intercept,'7.4f'))
30 print('rvalue= ',format(rvalue,'7.4f'))
31 print('pvalue= ',format(pvalue,'7.2e'))
32 print('se= ',format(se,'7.4f'))
33 # Plot the data
34 plt.plot(tData,PData,linestyle='',marker='d',label='Data')
35 plt.xlabel('T [K]')
36 plt.ylabel('P [kPa]')
37 plt.xlim((0,500))
38
39 # Plot the theory
40 tTheory = np.linspace(0,500,100)
41 PTheory = slope*tTheory + intercept
42 plt.plot(tTheory,PTheory,label='Model')
43 plt.legend()
44 plt.savefig('GasData.png')
```

Figure 10.4.: Python code to fit gas data to linear model.

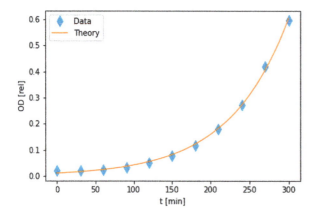

Figure 10.5.: Growth of bacteria during the exponential growth phase.

10.2.3. Non-linear Regression

10.2.3.1. Case 1: Model Function is Known Explicitly

The method of least-squares can also be applied to fitting data to a non-linear model, such as the exponential growth model. Figure 10.5 shows the growth of bacteria.

It is clear that a linear model will not describe this data. Indeed we know that an exponential growth model should work well. Equation 10.7 gives the equation defining this model.

$$N\left(t\right) = N_0 e^{rt} \tag{10.7}$$

The complication with applying the method of least-squares to a non-linear model is that we do not have explicit formulae for the model parameters, N_0 and r, as we did in the linear case. The computational problem is to perform a search through the parameter space to find the values that will minimize the Least Square Estimator function. This will work if we can help the computer out by giving a good starting place. We will make use of a Python library function that will search for the model parameter values that minimize the Least Squares Estimator function given an initial starting guess for the values. The function is from the scipy.optimize library. It is the scipy.optimize.curve_fit()function (The SciPy Community, 2022a). Here is the basic syntax for using the curve_fit function.

```
popt, pcov = scipy.optimize.curve_fit( f , tData , fData , p )
```

f is the callable function that calculates the theoretical value for the independent variable t. It must be in the form

$f(t, parameters)$

where t, the independent variable must be listed first, and parameters must be listed separately, not in a tuple variable.

tData and fData are lists containing independent and dependent variable values for data points.

popt will be the list of optimized parameter values.

pcov contains the covariance matrix. It gives information about the uncertainty in the parameter estimates. Explicit values for statistical uncertainty are obtained using

psterr = np.sqrt(np.diag(pcov))

psterr is a list containing the standard error of each parameter estimate in the same order as in popt.

Figure 10.6 shows a program that can fit the exponential growth model of Equation 10.7 to the data of Figure 10.5.

```python
"""
Program: Non-linear Least Squares - Explicit Function Form
Author: C.D. Wentworth
Version: 10.25.2022.1
Summary: This program reads in bacterial growth data and fits
the data to the non-linear exponential growth model using the
scipy.optimize.curve_fit() function.
Version History:
        3.28.2017.1: base version
        10.25.2022.1: added parameter uncertainty calculation

"""

import numpy as np
import matplotlib.pylab as plt
import scipy.optimize as so

def N(t,N0,r):
    tmp = N0*np.exp(r*t)
    return tmp

# import data
cols = np.loadtxt('BacterialGrowthData.txt', skiprows=9)
tData = cols[:, 0]
ODData = cols[:, 1]

# Define initial guess for model parameters
N0 = 0.021
r = 0.020
p = N0,r

# Fit data to exponential growth model
popt,pcov = so.curve_fit(N, tData, ODData, p)
p_stderr = np.sqrt(np.diag(pcov))

# Plot the data
plt.plot(tData, ODData, linestyle='', marker='d',
         markersize=10.0, label='Data')
```

Figure 10.6a: Python program for using the curve_fit function.

```
40 # Plot the theory
41 N0,r = popt
42 dN0 = p_stderr[0]
43 dr = p_stderr[1]
44 tTheory = np.linspace(0, 300, 50)
45 ODTheory = N(tTheory,N0,r)
46 plt.plot(tTheory, ODTheory, label='Theory')
47 plt.xlabel('t [min]')
48 plt.ylabel('OD [rel]')
49 plt.legend()
50 plt.savefig('bacterialGrowthFitData.png')
51
52 # print out best-fit model parameters
53 print('N0 = ', format(N0,'6.4f'), '+-', format(dN0,'7.4f'))
54 print('r = ', format(r,'6.4f'), '+-', format(dr,'6.4f'))
```

Figure 10.6b: Continuation of Python program for using the curve_fit function.

Figure 10.5 shows the model fit to the data. The program will produce the following printed output giving least squares fit model parameter values and uncertainties.

N0 = 0.0112 +- 0.0007

r = 0.0133 +- 0.0002

10.2.3.2. Case 2: Model Function value must be obtained by numerical integration

We have worked with mathematical models where the actual model function describing how the dependent variable depends on the independent variable is not known, but instead the model is defined by an ordinary differential equation or system of such equations. We learned how to numerically integrate the differential equation to obtain model values for selected independent variable values.

How can we use the least squares technique for finding best-fit model parameters for this case where we must integrate a differential equation to get model values? We can still use the curve_fit function to perform the least squares minimization search, but we must do some extra work to give it the required callable model function. The function will be defined to have the form

f(t, parameters)

but inside this function definition we use the numerical integration function scipy.integrate.solve_ivp and integrate the differential equation from the initial independent variable value up to the desired independent variable value. The function will return just this last calculated dependent variable value, as required by the curve_fit function.

We will continue to work with the exponential growth model as a test case for working through the details of this method. The rate equation for the population function $y(t)$ is

$$\frac{dy}{dt} = ry(t) \tag{10.8}$$

and there is a known initial condition of the form

$$y(0) = y_i \tag{10.9}$$

Let's start with the basic setup required to use solv_ivp to integrate the exponential growth rate equation that we discussed back in Chapter 7. We must define a Python function f that calculates the rate of change of the state variable y for an array of independent variable values t.

```python
import scipy.integrate as si
import numpy as np
import matplotlib.pylab as plt
import scipy.optimize as so

# Create a function that defines the rhs of the differential equation
    system

def f(t,y,r):
#    y = a list that contains the system state
```

```
#    t = the time for which the right-hand-side of the system equations
#        is to be calculated.
#    r = a parameter needed for the model
#
#    Unpack the state of the system
     y0 = y[0] # value of the state variable y at t
#    Calculate the rates of change (the derivatives)
     dy0dt = r*y0
     return [dy0dt]
```

Next, we will define the callable model function required by curve_fit that will use the numerical integration method to calculate the value of the state variable y. We will name this function yf.

```
def yf(t,yi,r):
    p = r,
    yv = []
    y_0 = [yi]
    for tt in t:
        if np.abs(tt)<1.e-5:
            yvt = yi
        else:
            ta = np.linspace(0,tt,10)
            # Solve the DE
            sol = si.solve_ivp(f,(0,tt),y_0,t_eval=ta,args=p)
            yvt = sol.y[0][-1]
        yv.append(yvt)
    return yv
```

The callable function yf for the curve_fit function is designed to accept an array of independent variable values in the t array, which will normally be a numpy array. This requires the for loop to calculate the model function value by numerical integration for each value in t. The Python code that implements this idea is shown in Figure 10.7 and is contained in the file expGrowthModelNonLinRegNumSol.py.

```
1  """
2  Program Name: Non-linear Least Squares -
3               Numerical Integration Form
4  Author: C.D. Wentworth
5  version: 10.27.2022.1
6  Summary: This program reads in bacterial growth data and fits
7           the data to the non-linear exponential growth model
8           using the scipy.optimize.curve_fit() function. This
9           version uses a numerical integration method for
10          calculating values of the model function.
11 History:
12          3.17.2020.1: base
13          10.27.2022.1: changed variable names
14
15 """
16
17 import scipy.integrate as si
18 import numpy as np
19 import matplotlib.pylab as plt
20 import scipy.optimize as so
21
22 # Create a function that defines the rhs of the differential equation
        system
23
24 def f(t,y,r):
25 #    y = a list that contains the system state
26 #    t = the time for which the right-hand-side of the system equations
27 #        is to be calculated.
28 #    r = a parameter needed for the model
29 #
30
31 #    Unpack the state of the system
32      y0 = y[0] # value of the state variable at t
33
34 #    Calculate the rates of change (the derivatives)
35      dy0dt = r*y0
36
37      return [dy0dt]
38
39 def yf(t,yi,r):
40      p = r,
41      yv = []
42      y_0 = [yi]
43      for tt in t:
44          if np.abs(tt)<1.e-5:
45              yvt = yi
46          else:
47              ta = np.linspace(0,tt,10)
48              # Solve the DE
49              sol = si.solve_ivp(f,(0,tt),y_0,t_eval=ta,args=p)
50              yvt = sol.y[0][-1]
51          yv.append(yvt)
52      return yv
```

Figure 10.7a: Python code for performing nonlinear least squares data fitting using numerical integration.

```python
40  # Main Program
41
42  # Read in data
43  cols = np.loadtxt('BacterialGrowthData.txt',skiprows=9)
44  td = cols[:,0]
45  OD = cols[:,1]
46
47  # Define initial guess for model parameters
48  yi = 0.022
49  r = 0.02
50  p = (yi,r)
51
52  popt,pcov = so.curve_fit(yf,td,OD,p)
53  p_stderr = np.sqrt(np.diag(pcov))
54
55  # Calculate theoretical values
56  # Define the time grid
57  ta = np.linspace(0,300,200)
58  yi,r = popt
59  dyi = p_stderr[0]
60  dr = p_stderr[1]
61
62  yTheory = yf(ta,yi,r)
63
64  # print out best-fit model parameters
65  print('yi = ', format(yi,'6.4f'), '+-', format(dyi,'7.4f'))
66  print('r = ', format(r,'6.4f'), '+-', format(dr,'6.4f'))
67
68  # Plot the solution
69  plt.plot(td,OD,linestyle='',marker='o',markersize=10.0,
70          label='Data')
71  plt.plot(ta,yTheory,color='g', label='Theory')
72  plt.legend()
73  plt.xlabel('t [min]')
74  plt.ylabel('OD [rel]')
75  plt.savefig('bacterialGrowthFitDataNumInt.png', dpi=300)
76  plt.show()
```

Figure 10.7b: Continuation of Python code for performing nonlinear least squares data fitting using numerical integration.

The graph and print out of the best-fit model parameters are exactly the same as were produced by the code in Figure 10.6, which provides evidence that the more complicated numerical integration method for calculating the model function values works for the curve_fit function.

10.3. Computational Problem Solving: Damped Spring – Best-fit Model Parameter Values

We return now to the problem of the damped spring, described at the beginning of the chapter. The experimental setup is shown in Figure 10.1 and the data is contained in the file dampedSpringData.txt.

Problem Statement: Given the position-time data shown in Figure 10.2, find the best-fit model parameters by fitting the data to the damped spring position-time model function given by Equation (10.1) reproduced here.

$$y(t) = Ae^{-bt/2m} \cos(\omega t + \phi_0)$$

$$\omega = \sqrt{\omega_0^2 - \frac{b^2}{4m^2}} \; , \; \omega_0 = \sqrt{\frac{k}{m}}$$

(10.10)

10.3.1. Analysis: Analyze the problem

We will not reproduce the physics analysis of the situation here, since it is covered in standard university physics textbooks (Knight, 2016). The mass of the object is assumed to be fixed at the measured value of m = 0.550 [kg]. A static force constant value of k=22.9 [N/m] is also provided with the data. This value of k will not necessarily be the best-fit value for the damped oscillator. The model parameters that must be determined are (A, b, k). We will use the least squares technique to find the optimum values of these parameters. We will need to use the nonlinear version of this method available to us through the curve_fit function.

10.3.2. Design: Describe the data and develop algorithms

The position data shows an oscillation about a nonzero equilibrium point, but the model function we are using assumes an oscillation about $y=0$. We can adjust the position data by a constant amount to achieve this. Use the basic data plotting template and add the following line to adjust the position.

```
# Read in data
cols = np.loadtxt('dampedSpringData.txt',skiprows=3)
tData = cols[:,0]
```

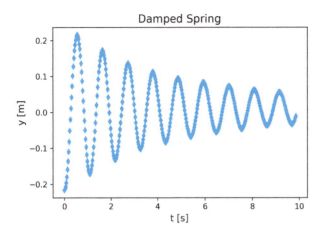

Figure 10.8.: Adjusted spring position data.

```
yData = cols[:,1]

# Adjust the y values by a constant amount
yData = yData - 0.7
```

An adjustment value of 0.7 appears to work well, as shown in Figure 10.8. The
adjustment is applied to all values in the array yData since it is a numpy array.

Starting guesses for the model parameters must be provided to the curve_fit function. Here are some possible values to use.

$$
\begin{aligned}
A &= 0.22 \,[\text{m}] \\
b &= 1.0 \\
k &= 22.9 \,[\text{N/m}] \\
\phi_0 &= \pi \,[\text{radians}]
\end{aligned}
$$

(10.11)

We can reuse the algorithm design used in section 10.2.3.1. We just need to replace
the model function N used in that code with the model function from Equation
10.10.

10.3.3. Implementation

The design described above implemented in Python code is shown in Figure 10.9. The code is in the file dampedSpringModelNonlinearRegression.py.

```python
1  """
2  Program: Damped Spring Model Data Fitting
3  Author: C.D. Wentworth
4  Version: 10.25.2022.1
5  Summary: This program reads in position-time data for
6  an object attached to a damped spring and fits
7  the data to the standard damped spring model function
8  using the scipy.optimize.curve_fit() function.
9  Version History:
10     10.25.2022.1: base
11
12 """
13
14 import numpy as np
15 import matplotlib.pylab as plt
16 import scipy.optimize as so
17
18 def yf(t, A, b, k, phi0):
19     m = 0.550  # [kg]
20     w0 = np.sqrt(k/m)
21     w = np.sqrt(w0**2 - b**2/(4.0*m**2))
22     tmp = A*np.exp(-b*t/(2.0*m))*np.cos(w*t + phi0)
23     return tmp
24
25 # Read in data
26 cols = np.loadtxt('dampedSpringData.txt',skiprows=3)
27 tData = cols[:,0]
28 yData = cols[:,1]
29
30 # Adjust the y data values by a constant amount
31 yData = yData - 0.7
32
33 # Define initial guess for model parameters
34 A = 0.22
35 b = 1.0
36 k = 22.9
37 phi0 = np.pi
38 p = A, b, k, phi0
```

Figure 10.9a: Python code for performing the least squares fit of spring position data to the damped spring model.

```
40  # Fit data to the damped spring model
41  popt,pcov = so.curve_fit(yf, tData, yData, p)
42  p_stderr = np.sqrt(np.diag(pcov))
43
44  # Plot the data
45  plt.plot(tData, yData, linestyle='', marker='d',
46          markersize=7.0, label='Data')
47
48  # Plot the theory
49  A, b, k, phi0 = popt
50  dA = p_stderr[0]
51  db = p_stderr[1]
52  dk = p_stderr[2]
53  dphi0 = p_stderr[3]
54  tTheory = np.linspace(0, 10, 500)
55  yTheory = yf(tTheory, A, b, k, phi0)
56  plt.plot(tTheory, yTheory, color='red', linewidth=3, label='Theory')
57  plt.xlabel('t [s]')
58  plt.ylabel('y [m]')
59  plt.legend()
60  plt.title('Least Squares Fit to Damped Spring Model', fontsize=14)
61  plt.savefig('dampedSpringFitData.png', dpi=300)
62
63  # print out best-fit model parameters
64  print('A = ', format(A,'6.4f'), '+-', format(dA,'7.4f'))
65  print('b = ', format(b,'6.4f'), '+-', format(db,'6.4f'))
66  print('k = ', format(k,'6.3f'), '+-', format(dk,'6.3f'))
67  print('phi0 = ', format(phi0,'6.3f'), '+-', format(dphi0,'6.3f'))
```

Figure 10.9b: Continuation of Python code for performing the least squares fit of spring position data to the damped spring model.

10.3.4. Testing

The code in Figure 10.9 produces the following print and graphical output.

```
A =   0.2203 +-   0.0024
b =   0.2084 +- 0.0038
k =  18.662 +-   0.023
phi0 =   3.072 +-  0.011
```

Figure 10.10 shows a reasonable fit between the position data and the theoretical model, although there may be some bias in the theory towards lower position values. This might be due to the constant adjustment made on the position data to yield oscillations about $y = 0$. This could be easily tested by adjusting the constant used in line 31.

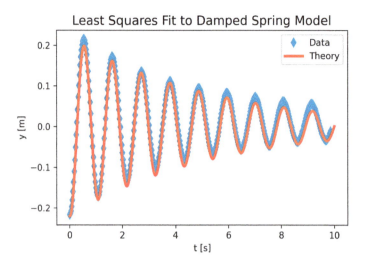

Figure 10.10.: Graph of damped spring data and model function with best-fit parameters.

10.4. Exercises

1. Which statements below are true for the Least Squares Method for fitting data to a model?

   a. It only applies to a model with one independent variable.

   b. It allows more than one model parameter to be estimated at a time.

   c. It involves minimizing the Least Squares Estimator Function.

   d. It can only be used to fit data to a linear model.

2. What is the calculus rule used to derive the least squares formulae for slope and intercept when fitting data to a linear model?

3. What is the name of the Python function that can perform a least squares fit of data to linear model?

4. What is the name of the Python function that can perform a least squares fit of data to nonlinear model?

10.5. Program Modification Problems

1. The Python program contained in Ch10ProgModProb1.py performs a linear regression (least-squares) fitting of some pressure versus temperature data for a gas held at constant volume. The data file for this program is GasData.txt.

The data file irisData.txt gives petal length and petal width measurements for many species of flower. You need to modify the Python code so that it performs a linear regression on this data. The program should also produce a plot of the data and the best-fit model. The graph should have

- proper axis labels
- a legend that shows what is the model and what is the data
- major gridlines

The program should print out a message that gives

- best-fit values for model parameters
- standard error for the slope
- the R^2 value
- the p-value for the fit

2. The Python program contained in Ch10ProgModProb2.py performs a non-linear regression (least-squares) fitting of bacterial growth data to the exponential model. The data file for this program is bacterialGrowthData.txt . The program fits the exponential model to the bacteria grown in Cold Spring Harbor A + 0.1% glucose (CSHA) media. The file V_natriegensGrowthData.txt contains growth data for another bacteria species. You need to modify the Python code so that it performs a

non-linear regression on this data using the logistics model function. The theoretical model function is

$$y(t) = \frac{My(0)e^{rt}}{M + y(0)(e^{rt} - 1)} \tag{10.12}$$

You can use the initial OD reading for $y(0)$, and then find the best-fit model parameters M and r. The program should also produce a plot of the data and the best-fit model. The graph should have

- proper axis labels
- a legend that shows what is the model and what is the data
- major gridlines

The program should print out

- best-fit values for the model parameters
- the standard error for the model parameter estimates

3. The Python program contained in Ch10ProgModProb3.py performs a non-linear regression (least-squares) fitting of bacterial growth data to the exponential model using the numerical integration technique for obtaining the model function. The data file for this program is bacterialGrowthData.txt . The program fits the exponential model to the bacteria grown in Cold Spring Harbor A + 0.1% glucose (CSHA) media. The file V_natriegensGrowthData.txt contains growth data for another bacteria species. You need to modify the Python code so that it performs a non-linear regression on this data using the logistics model function obtained through the numerical integration technique. The theoretical rate equation is

$$\frac{dy(t)}{dt} = ry\left(1 - \frac{y}{M}\right) \tag{10.13}$$

You can use the initial OD reading for $y(0)$, and then find the best-fit model parameters M and r using the least squares technique. The program should also produce a plot of the data and the best-fit model. The graph should have

■ proper axis labels
■ a legend that shows what is the model and what is the data
■ major gridlines

The program should print out

■ best-fit values for the model parameters
■ the standard error for the model parameter estimates

You should get the same final results as found in the Chapter 10 Program Modification Problem 2.

10.6. Program Development Problems

1. The following video shows a physics experiment where six bundled coffee filters are dropped from rest.

https://youtu.be/muxKA3kK4IA

The mass of the six filters is mass = 5.3 [g]. The vertical position of the filters as a function of time was obtained from the video using video analysis software. The data is contained in the file coffeeFilterData.txt.

Your task is to develop a dynamical systems model for the coffee filter motion. Go through the steps of our general problem-solving strategy. Develop the model and implement its solution in Python code. Estimate any model parameters using least squares fitting of the model to the data. Include an estimate of uncertainty in the model parameter(s). You can assume the mass of the filters and the acceleration due to gravity are already known, so the only model parameter that must be estimated is the air drag coefficient, k. An air drag force proportional to the square of the velocity will probably work well for this problem.

The position data starts at a nonzero time value. It may simplify developing the model to adjust the times to start at zero by subtracting the constant initial time from each time value.

Add a plot of the constant acceleration model (just gravity) to the figure so that a comparison can be made between the simple theory and the more complex one that includes air drag.

Create a documentation essay that describes how you went through the problem-solving strategy, detailing each step. Make sure you include a plot of the data and the best fit to the model. Specify the best-fit values of the model parameters with uncertainty.

10.7. References

Akritas, M. (2018). Chapter 6 Fitting Models to Data. In *Probability & Statistics with R for Engineers and Scientists (Classic Version) (Pearson Modern Classics for Advanced Statistics Series)*. Pearson.
http://www.librarything.com/work/17953304/book/228047240

Knight, R. (2016). Chapter 15 Oscillations. In *Physics for Scientists and Engineers: A Strategic Approach with Modern Physics* (4th edition). Pearson.

The SciPy Community. (2022a). *scipy.optimize.curve_fit—SciPy v1.9.3 Manual*.
https://docs.scipy.org/doc/scipy/reference/generated/scipy.optimize.curve_-fit.html

The SciPy Community. (2022b). *scipy.stats.linregress—SciPy v1.9.3 Manual.* https://docs.scipy.org/doc/scipy/reference/generated/scipy.stats.linregress.html

11. Project Ideas: Dynamical Systems Models

11.1. Overview

One of the best ways to develop computational science skills for creating and solving models of scientific and engineering systems is to work on substantive projects that require analysis of the problem, finding appropriate data, and then developing code by reusing elements learned in previous chapters.

In this chapter, we will focus on developing dynamical systems models of several topics in the areas of biology, engineering, and physics. For each topic, we will present some basic background, provide references for additional background, define some of the key features of a model, and then suggest questions that can be explored with a computational solution to the model.

11.2. Biology

11.2.1. HIV Infection of an Individual

Human Immunodeficiency Viruses (HIV) are virus species that cause Acquired Immunodeficiency Syndrome (AIDS). There are two known species of HIV, labeled HIV-1 and HIV-2, both of which can cause AIDS. HIV are retroviruses, which contain RNA that can produce the DNA required for viral reproduction once a virus particle infects a host cell. HIV infects primarily T-lymphocyte cells, although they eventually infect a wide variety of cell types in the human host (Levy, 1993).

Figure 11.1 shows the life cycle of viral load and CD4+ Lymphocyte concentration in an individual after infection (Wikipedia Contributors, 2022). The time development of the infection can be divided into three periods: the primary (or acute) infection, the latent (or asymptotic or chronic) period, and finally the onset of AIDS. The appearance of AIDS, usually several years after HIV infection, is associated with the destruction of the immune system, as illustrated by the significant

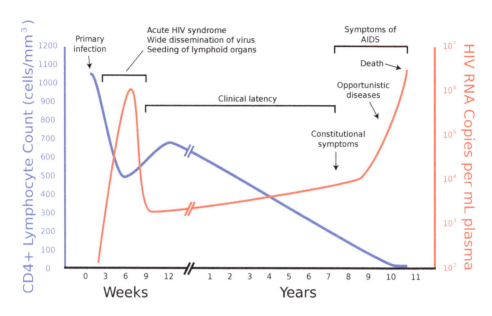

Figure 11.1.: Generalized view of virus particle and CD4+ Lymphocyte concentration after infection in an individual.

reduction in CD4+ T-cell concentration.

Mathematical modeling of HIV infection dynamics has played a critical role in understanding how AIDS develops and in aiding creation of effective antiviral treatments (Ribeiro & Perelson, 2004). In this project you will focus on developing a model of HIV dynamics during the acute stage of the infection. In particular, the model must predict the initial rapid increase in viral load and then its decrease to a stable longer term value. Additionally, the model should predict the initial rapid decrease in the uninfected CD4+ T-cell concentration and then its rise to a longer-term value that slowly decreases over time.

A good approach to this modeling problem is to use the dynamical systems framework with different compartments to represent the kind of cells and free viral particles that will be considered. The simplest compartment model that shows success at modeling the primary stage of infection is a three compartment model described by the state variables

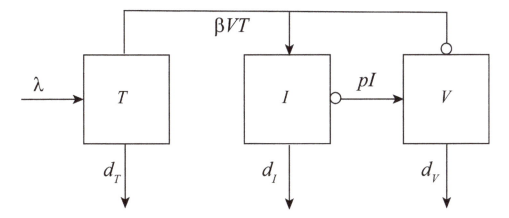

Figure 11.2.: Representation of the 3-compartment model for the HIV infection.

T = concentration of uninfected T-cells
I = concentration of infected T-cells
V = concentration of free virus

and represented by Figure 11.2.

The rate equations associated with Figure 11.2 are (Perelson & Ribeiro, 2013)

$$\frac{dT}{dt} = \lambda - d_T T - \beta V T$$

$$\frac{dI}{dt} = \beta V T - d_I I \qquad (11.1)$$

$$\frac{dV}{dt} = pI - d_V V$$

11.2.2. Epidemiology of the COVID-19 Infection

Epidemiology is the study of patterns in the development of a disease in a population, and is of interest to researchers in basic science and by public health policymakers. One disease that is of particular interest at the time of publication is COVID-19, a serious global pandemic that appeared at the end of 2019.

COVID-19 is a disease caused by the severe acute respiratory syndrome coronavirus 2 (SARS-CoV-2). There have been 651,918,402 confirmed cases and 6,656,601 deaths globally of COVID-19 since December 2019 (*WHO Coronavirus (COVID-19) Dashboard*, 2020). In addition to the clear health consequences, the disease triggered significant economic and political disruption throughout the world.

Mathematical models of epidemics in a population can aid exploration of basic questions about an infectious disease and offer guidance to policy makers about preparing for or responding to an epidemic. These observations are true for the study of the COVID-19 pandemic (Adiga et al., 2020). Mathematical models can help answer critical questions such as

1. How quickly will the disease sweep through a population?
2. How many people will be infected during the outbreak?
3. Will the disease persist in the population?

The starting point for investigating population-level dynamics of infectious diseases is the Susceptible-Infectious-Recovered (SIR) compartmental model. The state variables are

$$S = \text{The number of susceptible people}$$
$$I = \text{The number of infected people}$$
$$R = \text{The number of recovered people}$$

The basic SIR model is represented in Figure 11.3. The assumptions that are used in the basic model are

1. There is no birth and death of people, so the total population of people is a constant.
2. Susceptible people will become infected through exposure to infected people.
3. Infected people are immediately infectious.
4. The rate of recovery γ is assumed to be constant and recovered people are immune to infection.

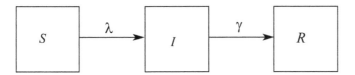

Figure 11.3.: Graphical representation of the basic SIR model.

The rate equations for the state variables that describe the model shown in Figure 11.3 are

$$\frac{dS}{dt} = -\lambda(I)S$$

$$\frac{dI}{dt} = \lambda(I)S - \gamma I \tag{11.2}$$

$$\frac{dR}{dt} = \gamma I$$

Since the total population N is constant we have the following relationship between S, I, and R.

$$N = S + I + R \tag{11.3}$$

Therefore, the dynamics of the system are really determined by two rate equations rather than three. The function $\lambda(I)$ is called the force of infection and gives the per capita rate at which susceptible individuals acquire infection. Different versions of the SIR model can be defined by choosing different forms for $\lambda(I)$. A typical choice is given by

$$\lambda(I) = \beta\frac{I}{N} \tag{11.4}$$

β is called the transmission rate.

We can convert the state variables to dimensionless form by dividing by the total population.

$$s = \frac{S}{N} \; , i = \frac{I}{N} \; , r = \frac{R}{N} \tag{11.5}$$

With these definitions for s, i, and r the rate equations become

$$\frac{ds}{dt} = -\beta si$$

$$\frac{di}{dt} = \beta si - \gamma i \tag{11.6}$$

$$\frac{dr}{dt} = \gamma i$$

and the conservation law Equation 11.3 becomes

$$1 = s + i + r \tag{11.7}$$

Figure 11.4 shows an example of the results of this model. This model shows an infection that always dies out eventually with all infected people eventually recovering.

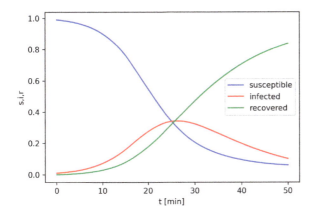

Figure 11.4.: Solution of the basic SIR model with $\beta = 0.30$ and $\gamma = 0.09$.

11.3. Physics

11.3.1. Bungee Jumping

Bungee jumping involves attaching a long bungee cord, a type of elastic cord, to a person and then having the person step off a bridge and plunge downwards, and then be pulled back up by the cord, shown in Figure 11.5 (Spy007au, 1996).

In the modern world it is a popular form of extreme activity, but there are examples of similar activities in history including the land diving ritual performed by men living on Pentecost Island, Vanuatu, shown in Figure 11.6 (Stein, 2010).

We will develop a basic mathematical model for the trajectory of a jumper using Newton's Second Law (Menz, 1993). The coordinate system that will be used to describe the jumper's position is shown in Figure 11.7. The origin of the y-axis is at the platform where the jumper steps off, and we take the +y direction to be down. The unstretched length of the cord is L. The mass of the jumper is m. We assume the bungee cord is massless.

The y-component of Newton's Second Law is

Figure 11.5.: Bungee jumping off the Zambezi Bridge, Victoria Falls, Africa (Spy007au, 1996).

Figure 11.6.: Land diving on Pentecost Island (Stein, 2010).

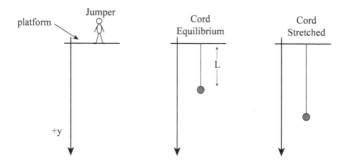

Figure 11.7.: Coordinate system for the bungee jump.

$$\sum_i F_{iy} = m\ddot{y}$$

$$mg - B(t) H(y - L) - D(t) = m\ddot{y} \tag{11.8}$$

which gives

$$\ddot{y} = g - \frac{1}{m}B(t) H(y - L) - \frac{1}{m}D(t) \tag{11.9}$$

Note that the bungee force is nonzero only when $y>L$. This fact can be implemented by using the Heaviside function

$$H(y) = \begin{cases} 1\, , y \geq 0 \\ 0\, , y < 0 \end{cases} \tag{11.10}$$

The following model for the air drag force will be used.

$$D_y(v_y) = -\frac{1}{2}\rho C_D A\left(v_y^2\right) sign(v_y) \tag{11.11}$$

where

$\rho = $ density of air
$C_D = $ dimensionless drag coefficient for the rocket
$A = $ cross-sectional area of the rocket
$$sign(v_y) = \begin{cases} +1\, , v_y > 0 \\ -1\, , v_y < 0 \end{cases}$$

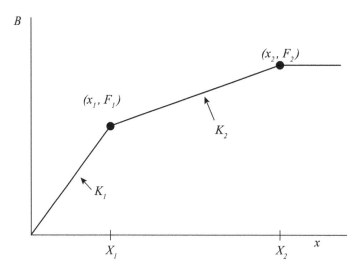

Figure 11.8.: Bungee force model.

We need a model for the bungee force B. The simplest bungee force model is to assume that it acts as an ideal spring

$$B = k_1 (y - L) \tag{11.12}$$

A more realistic bungee force is suggested by Figure 11.8 (Menz, 1993)

where x is the amount of cord stretch from equilibrium. The equation describing the force implied by this figure is

$$B = \begin{cases} K_1 x \, , 0 \leq x \leq X_1 \\ K_2 x + (K_1 - K_2) X_1 \, , X_1 < x \leq X_2 \\ K_2 X_2 + (K_1 - K_2) X_1 \, , X_2 < x \end{cases} \tag{11.13}$$

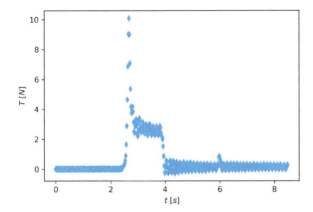

Figure 11.9.: Thrust data for the Estes B4-2 engine.

11.3.2. Model Rocket Trajectory

Model rocketry is a popular hobby for many adults and children. The rockets are constructed from cardboard, plastic, and balsa wood and are powered by commercially available solid propellant engines that can generate a thrust force of a few Newtons over one to three seconds (*Estes Rockets*, 2021). Figure 11.9 shows an example of the thrust data for a particular engine. The second small thrust peak corresponds to the parachute ejection shot. It can be ignored in the trajectory analysis that follows.

The trajectory of the model rocket can be predicted from the thrust data by performing a Newton's Second Law analysis. The resulting equations of motion will be most easily studied using a numerical integration procedure, as discussed in previous chapters. The Newton's 2nd Law analysis begins with a free body diagram. The physical forces acting on the rocket include gravity, $\overrightarrow{\mathbf{F}}_g$, the rocket thrust, $\overrightarrow{\mathbf{T}}$, and the air drag, $\overrightarrow{\mathbf{D}}$. Figure 11.10 shows the free body diagram and the definition of the +y direction that will be assumed in the analysis.

We introduce the following function definitions

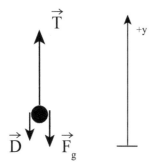

Figure 11.10.: Model rocket free body diagram while rocket is not in contact with the ground.

$m\left(t\right)$ = mass of the rocket and engine as a function of time t
$D\left(t\right)$ = magnitude of the drag force as a function of time
$T\left(t\right)$ = magnitude of the thrust force as a function of time

Newton's Second Law expressed in terms of momentum is

$$\overrightarrow{F}_{net} = \frac{d\overrightarrow{p}}{dt}, \overrightarrow{p} = m\left(t\right)\overrightarrow{v} \tag{11.14}$$

We will assume that the rocket travels only in the vertical direction, so only the y-component of the forces and Newton's Second Law will be needed.

$$F_{gy} = -m\left(t\right)g \tag{11.15}$$

The model for the air drag force is given by Equation 11.11. We also need a model for the thrust function based on the available thrust data. This will be developed below.

Newton's Second Law, Equation , can now be written as

$$T\left(t\right) + D_y\left(v_y\right) - m\left(t\right)g = \frac{dm}{dt}v_y\left(t\right) + m\left(t\right)a_y\left(t\right) \tag{11.16}$$

where

$$a_y\left(t\right) = \text{y-component of acceleration at time t}$$

We will assume that the mass of the rocket will not change as the engine burns since the fuel mass is a very small percentage of the total mass. This assumption gives

$$\frac{dm}{dt} = 0 \tag{11.17}$$

which allows us to write Equation 11.16 as

$$a_y\left(t\right) = \frac{1}{m}\left[T\left(t\right) + D_y\left(v_y\right) - mg\right] \tag{11.18}$$

Next, we will develop a mathematical model for the rocket thrust. Figure 11.9 suggests that a possible mathematical model for the thrust is given by Figure 11.11.

Using analytical geometry, we can find the equation representing Figure 11.11 to be

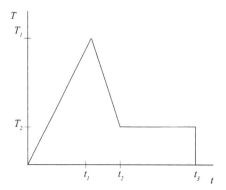

Figure 11.11.: Graphical representation of the rocket thrust model.

$$T(t) = \begin{cases} s_1 t \, , 0 \leq t \leq t_1 \\ s_2 t + b \, , t_1 < t \leq t_2 \\ T_2 \, , t_2 < t \leq t_3 \\ 0 \, , t > t_3 \end{cases} \qquad (11.19)$$

where

$$s_1 = \frac{T_1}{t_1} \, , s_2 = \frac{(T_2 - T_1)}{(t_2 - t_1)} \, , b = (T_1 - s_2 t_1)$$

Additional background information about model rocket trajectories can be found in (Keeports, 1990).

11.4. Projects

HIV Infection of an Individual

1. Fit the data in the file HIVData.txt to the HIV model defined by Equation 11.1 using the least squares method to find the best fit model parameters. Explore how making variations in model parameters affects the time behavior of the infection.
2. Explore the effect of antiretroviral therapy by modifying the rate equation for V as suggested by (Perelson & Ribeiro, 2013). First explore how the theoretical predictions for the original model of Equation 11.1 compare with theoretical predictions of the modified model. Next, locate data on viral concentration after antiretroviral therapy and compare the viral clearance rate, d_V, with therapy and without therapy.

Epidemiology of the COVID-19 Infection

1. Use the data file owid-covid-data.xlsx from the Our World in Data website (Mathieu et al., 2020) to obtain COVID infection data for the US. Fit the US data for the early stage of the epidemic to the model of Equation 11.2. Is there any time period for which the basic SIR model provides an adequate description? Calculate the basic reproductive rate for the infected population (see (Blackwood & Childs, 2018) for discussion of the reproductive rate).
2. Add birth and death terms to the rate equations. Implement a numerical solution to the revised model. Explore the effect of changing birth and death rates on the time behavior of the model. What is the effect of birth and death on the basic reproduction rate?

Bungee Jumping

1. Implement a numerical solution to the model of Equation 11.9. Investigate bungee jumping sites to determine typical drop lengths for participants. If possible, determine typical times for a jumper to come to rest. Use this data to determine best fit model parameter values. Explore the behavior of velocity and acceleration over the time of the jump. Use the model to explore possible health effects of bungee jumping.
2. Implement a numerical solution to the model of Equation 11.9. Use the model to assist in designing a bungee jump site. You will need to decide about the range of jumper masses your site will accommodate, the maxi-

mum jump height, and the required bungee cords that will serve people of different mass. Research safety and health concerns and verify that your design meets all required safety criteria.

Model Rocket Trajectory

1. Use the thrust data given for the Estes A8-3, B4-2, and B6-2 engines to estimate the parameters for the thrust model given by Equation 11.19. The thrust data is contained in the files A8-3_ThrustData.txt, B4-2_Thrust-Data.txt, and B6-2_ThrustData.txt. Create a Python thrust force function for each of the engines. Choose a model rocket from the Estes catalog. Note the rocket's mass, cross-sectional area, and suggested maximum height. Implement a numerical solution of the trajectory model given by Equation 11.18 for each of the engines. Compare the theoretical maximum height with the height suggested by Estes.
2. Develop a mathematical model for a two-dimensional rocket trajectory. This will require analyzing motion in both the vertical and horizontal directions. Explore how the launch angle affects the maximum height and range of the rocket for each of the engines.

11.5. References

Adiga, A., Dubhashi, D., Lewis, B., Marathe, M., Venkatramanan, S., & Vullikanti, A. (2020). Mathematical Models for COVID-19 Pandemic: A Comparative Analysis. *Journal of the Indian Institute of Science*, *100*(4), 793–807. https://doi.org/10.1007/s41745-020-00200-6

Blackwood, J., & Childs, L. (2018). An introduction to compartmental modeling for the budding infectious disease modeler. *Letters in Biomathematics*, *5*(1), Article 1. https://doi.org/10.30707/LiB5.1Blackwood

Estes Rockets. (2021). Estes Industries. https://estesrockets.com/

Keeports, D. (1990). Numerical calculation of model rocket trajectories. *The Physics Teacher, 28*(5), 274–280. https://doi.org/10.1119/1.2343024

Levy, J. A. (1993). Pathogenesis of human immunodeficiency virus infection. *Microbiological Reviews, 57*(1), 183–289. https://doi.org/10.1128/mr.57.1.183-289.1993

Mathieu, E., Ritchie, H., Rodés-Guirao, L., Appel, C., Giattino, C., Hasell, J., Macdonald, B., Dattani, S., Beltekian, D., Ortiz-Ospina, E., & Roser, M. (2020). Coronavirus Pandemic (COVID-19). *Our World in Data.* https://ourworldindata.org/coronavirus-source-data

Menz, P. G. (1993). The physics of bungee jumping. *The Physics Teacher, 31*(8), 483–487. https://doi.org/10.1119/1.2343852

Perelson, A. S., & Ribeiro, R. M. (2013). Modeling the within-host dynamics of HIV infection. *BMC Biology, 11*, 96. https://doi.org/10.1186/1741-7007-11-96

Ribeiro, R. M., & Perelson, A. S. (2004). The Analysis of HIV Dynamics Using Mathematical Models. In *AIDS and Other Manifestations of HIV Infection* (4th ed.). Elsevier.

Spy007au. (1996). *Bungee jumping off the Zambezi Bridge, Victoria Falls, Africa.* Own work by the original uploader. https://commons.wikimedia.org/wiki/File:Bill%27s_Bungy_Jump.jpg

Stein, P. (2010). *The Tower.* https://commons.wikimedia.org/w/index.php?curid=10438782

WHO Coronavirus (COVID-19) Dashboard. (2020). WHO Coronavirus (COVID-19) Dashboard. https://covid19.who.int

Wikipedia Contributors. (2022). HIV. In *Wikipedia.*

https://en.wikipedia.org/w/index.php?title=HIV&oldid=1119384291

12. Project Ideas: Stochastic Models

12.1. Forest Fire Propagation

When forest fires encroach on regions with human habitation, the damage done can be substantial, including property damage, health effects, and death. Between 2017 and 2021 the average cost per year of wildfires in the U.S. was $16.8 billion, and the average number of deaths per year was 43(Smith, 2020). If health effects of particulate inhalation are included, the death toll from wildfires climbs to about 33000 deaths per year globally (Verzoni, 2021). Figure 12.1 shows the acres burned by wildfires in the U.S. between 1983 and 2021, which suggests an upward trend over this time (NICC, 2022).

As the acres burned and human migration into forest increases, the money spent on wildfire suppression also increases, as shown in Figure 12.2 (NIFC, 2022).

The economic and health costs of wildfires have provided motivation for developing models that can help us understand and predict forest fire propagation. Historically, fire propagation models were based on partial differential equations, but more recently cellular automata models have been developed (Perry, 1998). We will explore the use of cellular automata (CA) to model forest fire propagation since the technique is mathematically and computationally simpler than using PDE's.

Several cellular automata models for forest fires have been proposed. Some models focus on basic physics questions concerning self-organized criticality and are not meant to model real forest fires in detail (Bak et al., 1990). Other models attempt to build in features from observations of actual fires including effects of fuel moisture, wind speed, local temperature, and relative humidity (Clarke et al., 1994). More recent CA models incorporate real forest fire data to calculate the transition rules governing a cell's change of state (Collin et al., 2011) or incorporate the effects of inhomogeneous terrain (Encinas et al., 2007).

We will outline a basic forest fire cellular automaton model. Chapter 9 gives the

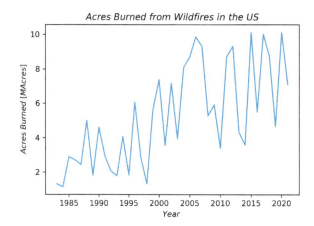

Figure 12.1.: Acres burned from forest fires in the U.S. (NICC, 2022).

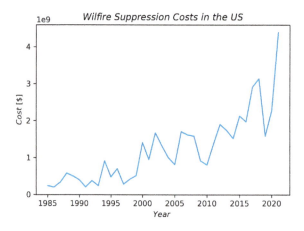

Figure 12.2.: . Federal wildfire suppression costs in the U.S. (NIFC, 2022).

basic elements of a cellular automaton model. It requires

- a discrete lattice of sites or cells in d dimensions, L
- operating using discrete time steps
- a discrete set of possible cell values, S
- a defined neighborhood, N of each cell in L
- a transition rule, Φ, that specifies how the state of a cell will be updated depending on its current value and the values of sites in the neighborhood.

The cellular automaton model is described by the collection $\langle L, S, N, \Phi \rangle$.

We will develop an adaptation of the Encinas model (Encinas et al., 2007). Our model will use a 2-dimensional lattice L that can represent the earth's surface. We will assume that L is square, with the number of grid points on one side set to a fixed value. A time step in the simulation corresponds to allowing each grid point in the lattice the chance to change state. The state of a cell is defined as

$$S_{ij} = \frac{A_{bij}}{A} \tag{12.1}$$

where A_{bij} is the burned area of the cell, and A is the total area of a cell. The neighborhood of a cell at (i,j) will be taken to be the Moore neighborhood illustrated in Figure 12.3 (Wikipedia Contributors, 2022). If (i,j) is on the lattice boundary then $S_{ij} = 0$.

The transition rule is

$$\Phi : S_{ij}^{(t+1)} = \frac{R_{ij}}{R} S_{ij}^{(t)} + \sum_{(\alpha,\beta) \varepsilon N_{adj}} \mu_{\alpha\beta} \frac{R_{i+\alpha,j+\beta}}{R} S_{i+\alpha,j+\beta}^{(t)}$$
$$+ \sum_{(\alpha,\beta) \varepsilon N_{diag}} \mu_{\alpha\beta} \frac{\pi(R_{i+\alpha,j+\beta})^2}{4R^2} S_{i+\alpha,j+\beta}^{(t)} \tag{12.2}$$

where $\mu_{\alpha\beta}$ incorporates physical effects including wind speed and slope height at

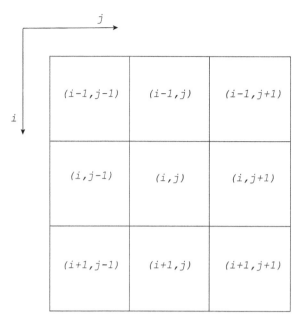

Figure 12.3.: Moore neighborhood for a 2-d cellular automaton.

each cell. The values (α, β) that define the neighbors of (i, j) in Equation 12.2 are given by

$$N_{adj} = \{(-1, 0), (0, 1), (1, 0), (0, -1)\} \tag{12.3}$$

$$N_{diag} = \{(-1, 1), (1, 1), (1, -1), (-1, -1)\} \tag{12.4}$$

In this version of the model, we will take $\mu_{\alpha\beta}$ to be

$$\mu_{\alpha\beta} = w_{i+\alpha, j+\beta} \cdot h_{i+\alpha, j+\beta} \tag{12.5}$$

and $w_{i+\alpha, j+\beta}$ are the entries in the wind speed matrix

$$W_{ij} = \begin{bmatrix} w_{i-1,j-1} & w_{i-1,j} & w_{i-1,j+1} \\ w_{i,j-1} & 1 & w_{i,j+1} \\ w_{i+1,j-1} & w_{i+1,j} & w_{i+1,j+1} \end{bmatrix} \tag{12.6}$$

and $h_{i+\alpha,j+\beta}$ are entries in the slope height matrix

$$H_{ij} = \begin{bmatrix} h_{i-1,j-1} & h_{i-1,j} & h_{i-1,j+1} \\ h_{i,j-1} & 1 & h_{i,j+1} \\ h_{i+1,j-1} & h_{i+1,j} & h_{i+1,j+1} \end{bmatrix} \tag{12.7}$$

If Equation 12.2 yields a value greater than 1 then $S_{ij}^{(t+1)}$ should be replaced by 1. Finally, to introduce a stochastic element into the state transition rule we assume that the cell state changes according to Equation 12.2 with probability p, otherwise it remains unchanged.

The state of a cellular automaton cell is usually taken to be discrete. Since Equation 12.2 can generally yield a continuous result, forming discrete values requires an additional step. One way to achieve state values that are discrete is to use the following definition. Assume the number of discrete values desired is N_s.

$$S'_{ij} = \begin{cases} 0 \, , \, S_{ij} = 0 \\ \frac{1}{(N_s-1)} \, , \, 0 < S_{ij} < \frac{1}{(N_s-1)} \\ \frac{2}{(N_s-1)} \, , \, \frac{1}{(N_s-1)} \leq S_{ij} < \frac{2}{(N_s-1)} \\ \vdots \\ 1 \, , \, S_{ij} \geq 1 \end{cases} \tag{12.8}$$

Some system properties that might be studied using the simulation include the percentage of cells that have not burned ($S_{ij} = 0$), are burning ($0 < S_{ij} < 1$), and have burnt out ($S_{ij} = 1$) as functions of time.

12.2. Solid-State Diffusion

Diffusion is the transport of matter through the movement of individual atoms or molecules. Macroscopically, diffusion involves movement from a region of high concentration to low concentration of the diffusing material. Diffusion can be contrasted with fluid flow, where all the atoms move together due to a pressure gradient. Solid-state diffusion involves the motion of individual atoms in a solid material.

Solid-state diffusion plays a significant role in many industrial applications. Examples of materials in which solid-state diffusion plays a significant role are steel, doped semiconductors, and superionic conductors.

Steel is made from iron by the addition of carbon to iron. Uniform distribution of carbon can be achieved by diffusion of carbon atoms during a sintering process. The carbon content of solid steel can be adjusted by diffusion of carbon at the solid surface, a process called carburizing (Mandal, 2015). Other elements, such as titanium or nitrogen, can be added to iron in the same way as carbon to change the mechanical properties of the steel.

The electrical and optical properties of semiconductor materials can be adjusted by doping the base material, such as silicon, with other elements, such as gallium. This can be done by exposing the silicon to a gas containing the dopant and allowing it to diffuse into the silicon (Zant, 2014).

Superionic, or fast-ion, conductors are materials in which charge transport is achieved by moving ions rather than moving electrons, as occurs in metallic conductors (Beard, 2019). These materials are important in creating solid-state batteries, supercapacitors, and fuel cells, where solid electrolytes are required. Charge transport in these materials can be modeled by a hopping motion of ions in the solid with a preferred direction of motion established by an electric field.

We will develop a model of solid-state diffusion based on the idea of a lattice gas, in which particles that can move are placed in a lattice structure and can move

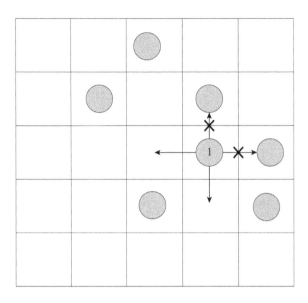

Figure 12.4.: Two-dimensional lattice gas.

between sites in the lattice. The following assumptions will be made for the model.

1. The diffusive motion is accomplished by a particle hopping to a nearby lattice site.
2. Particles can only hop to nearest-neighbor lattice sites, which are assumed to be separated by a distance a.
3. The particles do not interact with each other except that double occupancy of a lattice site is not allowed.
4. At each time step, a particle will choose a direction at random and attempt to move to the nearest-neighbor lattice site in that direction.

Figure 12.4 shows an example of a lattice gas in a two-dimensional, square lattice. The allowed nearest-neighbor hops for one particle are shown.

An important property that can be studied with this model is the tracer diffusion coefficient, D^*. Tracer diffusion refers to diffusion of a single particle that can be tracked over time. The tracer diffusion coefficient can be defined in terms of the mean square displacement of the tracer particle. For the square lattice, this definition is

$$D^* = \frac{|\vec{r}(t) - \vec{r}(0)|^2}{4t} \tag{12.9}$$

where $\vec{r}(t)$ is the position of the tracer at time t and the averaging, indicated by the angle brackets, is over all the diffusing particles in the lattice, since each one can be considered as a tracer.

We want to implement a Monte Carlo simulation of this model. The time t in a simulation will be considered as one pass through all the diffusing particles, attempting to allow each one to hop. The long-term time behavior of the mean square tracer displacement should be linear, giving a time-independent tracer diffusion coefficient. We can use the simulation to calculate D^* as a function of particle concentration in the lattice. Table 12.1 defines the most important data structures or variables that might be used in the simulation code.

12.3. Creating Animations with Python

Animations can be used to facilitate exploring a model or data particularly when identifying time-dependent patterns. Basic animations can be created using the matplotlib package. Two methods will be discussed here.

1. placing the matplotlib.pyplot pause function in a for loop
2. using the FuncAnimation function from the matplotlib.animation package

Animations can be useful in the early stages of analyzing the models discussed in this chapter.

Table 12.1.: Data structures used in the Monte Carlo simulation of diffusion

Variable Name	Data Type	Description
concentration	float	gives the fraction of lattice sites occupied by diffusing particles
init_particle_positions	list	init_particle_positions[p] is a list of the initial x/y coordinates of particle number p
lattice_size	integer	the length of one side of the lattice as a number of sites
num_of_particles	integer	the number of particles in the lattice
p	integer	particle number
particle_positions	list	particle_positions[p] is a list of the current x/y coordinates of particle number p
site_occupancy	2d numpy array of integers	site_occupancy[xi,yi] specifies whether the lattice site with coordinates (xi,yi) is occupied
t	integer	current time step being considered
total_time_steps	integer	the number of time steps in a simulation

```python
import matplotlib.pyplot as plt
import numpy as np

x = []
y = []

for i in range(100):
 x.append(i)
 y.append(np.sqrt(i))

 # Mention x and y limits to define their range
 plt.xlim(0, 100)
 plt.ylim(0, 10)

 # Ploting graph
 plt.plot(x, y, color = 'green')
 plt.pause(0.01)

plt.show()
```

Figure 12.5.: Example of animation using the pause function.

12.3.1. Using the pause() Function

The pause(interval) function from the matplotlib.pyplot package will pause program execution for the number of seconds indicated by the interval parameter. An animation can be created by setting up a for loop in which the plot function is executed and then the pause function is executed within the loop. Figure 12.5 shows an example of code that will animate plotting a function using the pause() function.

12.3.2. Using the FuncAnimation Function

The matplotlib.animation package has a function named FuncAnimation that requires a bit more work to set up than just using the pause function but allows for more flexibility including being able to save the animation in a gif, avi, mov, or mp4 file format. Using FuncAnimation to create an animation requires the following steps

1. defining an empty figure object
2. defining a callable Python function that can create one frame of the animation, usually performed using the matplotlib plot function
3. making a call to the FuncAnimation function

Defining the empty figure object is done using the Object Oriented interface for matplotlib. The basic syntax for creating an empty plotting grid is

```
fig = plt.figure()
axes = fig.add_subplot(1,1,1)
```

The callable function that creates a frame must have the frame number as the first argument with any other arguments following:

```
def func(frame, *fargs)
```

where frame will be an integer specifying the frame number being created.

FuncAnimation has many parameters (The Matplotlib development team, 2023). A basic call will be of the form

```
animation_name = animation.FuncAnimation(fig=fig_name,
                 func=frame_creation_function, interval=time ,
                 save_count=num_of_frames)
```

where

fig_name is the figure object

frame_creation_function is the callable function that creates one frame in the animation

interval is the time between frames in millisecond

num_of_frames is an integer that specifies the number of frames in the animation

animation_name is a user specified variable that refers to the animation object and can be used to play the animation or save it as video file

Figure 12.6 gives an example of a program that produces an animation of a two-dimensional random walk.

```
1  """
2  Title: Random Walk in 2D: animations
3  Author: C.D. Wentworth
4  version: 12.31.2022.1
5  Summary: This program performs a random walk on a
6           two-dimensional lattice. It uses reflective
7           boundary conditions. It produces an animation
8           of the walk
9  version history:
10         12.31.2022.1: base
11
12 """
13 import random as rn
14 import numpy as np
15 import matplotlib.pyplot as plt
16 from matplotlib import animation
17 from matplotlib import rc
18
19 rc('animation', html='html5')
20
21 def createFrame(t):
22     x_frame.append(x_array[t])
23     y_frame.append(y_array[t])
24     axes.set_title(str(t))
25     plt.plot(x_frame,y_frame, scaley=True, scalex=True, linewidth=2, color=
       "blue")
26
27
28 def step(xi, yi, lx, ux, ly, uy):
29     import random as rn
30     r = rn.randint(1, 4)
31     if r == 1:
32         # go east
33         if xi < ux:
34             xi = xi + 1
35     elif r == 2:
36         # go west
37         if xi > lx:
38             xi = xi - 1
39     elif r == 3:
40         # go north
41         if yi < uy:
42             yi = yi + 1
43     else:
44         # go south
45         if yi > ly:
46             yi = yi - 1
47     return xi, yi
```

Figure 12.6a: Example code that uses FuncAnimation to show a 2-d random walk.

```
50  # --Main Program
51  # set up random generator
52  rn.seed(42)
53
54  # define the grid
55  lx = -30
56  ux = 30
57  ly = -30
58  uy = 30
59
60  # set up the simulation
61  N = 500
62  xi = 0
63  yi = 0
64  position_array = np.zeros((N + 1, 2))
65
66  # execute random walk
67  for i in range(1, N + 1):
68      xi, yi = step(xi, yi, lx, ux, ly, uy)
69      position_array[i, 0] = xi
70      position_array[i, 1] = yi
71
72  x_array = position_array[:, 0]
73  y_array = position_array[:, 1]
74  x_frame = []
75  y_frame = []
76
77  # Create the animation
78  fig = plt.figure()
79  axes = fig.add_subplot(1,1,1)
80  axes.set_ylim(lx, ux)
81  axes.set_xlim(ly, uy)
82  #writervideo = animation.FFMpegWriter(fps=10)
83  ani = animation.FuncAnimation(fig=fig, func=createFrame, interval=100 ,
84                                save_count=N)
85  ani
```

Figure 12.6b: Continuation of Example code that uses FuncAnimation to show a 2-d random walk.

The animation object is named ani, and it is played in a Jupyter notebook by just entering the name as the last statement of the program. If you want to play the animation within an IDE such as Spyder then additional steps will usually need to be completed.

12.4. Projects

Forest Fire Propagation

Develop code that implements the model described in section 12.1. This can be done in stages, depending on available time. One way to start the fire is to initialize one cell to the state in the range $0 < S_{ij} < 1$. Another possible initial state would be to allow a given concentration c of the cells to be burning. For each of the following scenarios, measure the percentage of cells in the burning state as a function of time. Consider the effect of varying p and c.

1. Assume a homogeneous lattice, so there is no variation in burn rates, no wind, and no variation in height. This is accomplished by

■ setting all entries of R to one

■ letting the wind and height matrices be defined by

$$W_{ij} = \begin{bmatrix} 1 & 1 & 1 \\ 1 & 1 & 1 \\ 1 & 1 & 1 \end{bmatrix}, H_{ij} = \begin{bmatrix} 1 & 1 & 1 \\ 1 & 1 & 1 \\ 1 & 1 & 1 \end{bmatrix}$$

2. Consider an inhomogeneous system with burn rates that vary in some way on the lattice. You might consider dividing the lattice into two regions with different burn rates.

3. Explore the effect of wind speed. A north-to-south wind is represented by the wind matrix

$$W_{ij} = \begin{bmatrix} 1.5 & 1.5 & 1.5 \\ 1 & 1 & 1 \\ 0.5 & 0.5 & 0.5 \end{bmatrix}$$

Solid-State Diffusion

1. Implement the 2-dimensional diffusion model discussed in section 12.2. Study the mean square displacement of a particle $|\vec{r}(t) - \vec{r}(0)|^2$ as a function of time and as a function of concentration. Does it become linear after some period of time, indicating a well-defined tracer diffusion coefficient? Does the lattice size have an effect on the results? Remember that when performing a Monte Carlo simulation of a model, measured properties of the system will show random variation. To eliminate the effects of this random variation you must repeat the simulation multiple time and form a mean over all the simulations for any measured property, as discussed in Chapter 9.

2. Develop a model for diffusion in a one-dimensional lattice similar to the two-dimensional model. Implement a Monte Carlo simulation of the model. Does the mean-square displacement $|x(t) - x(0)|^2$ behave the same as in two dimensions?

12.5. References

Bak, P., Chen, K., & Tang, C. (1990). A forest-fire model and some thoughts on turbulence. *Physics Letters A, 147*(5), 297–300. https://doi.org/10.1016/0375-9601(90)90451-S

Beard, K. W. (Ed.). (2019). SECTION C: SOLID-STATE ELECTROLYTES (CERAMIC, GLASS, POLYMER). In *Linden's Handbook of Batteries* (5th edition.). McGraw-Hill Education. https://www.accessengineeringlibrary.com/content/book/9781260115925/toc-chapter/chapter22/section/section24

Clarke, K. C., Brass, J. A., & Riggan, P. J. (1994). A cellular automaton model of wildfire propagation and extinction. *Photogrammetric Engineering and Remote Sensing. 60(11): 1355-1367, 60*(11), Article 11.

Collin, A., Bernardin, D., & Séro-Guillaume, O. (2011). A Physical-Based Cellular

Automaton Model for Forest-Fire Propagation. *Combustion Science and Technology*, *183*(4), 347–369. https://doi.org/10.1080/00102202.2010.508476

Encinas, A. H., Encinas, L. H., White, S. H., Rey, A. M. del, & Sánchez, G. R. (2007). Simulation of forest fire fronts using cellular automata. *Advances in Engineering Software*, *38*(6), 372–378. https://doi.org/10.1016/j.advengsoft.2006.09.002

Mandal, S. K. (2015). Heat Treatment and Welding of Steels. In *Steel Metallurgy: Properties, Specifications and Applications* (First edition.). McGraw-Hill Education. https://www.accessengineeringlibrary.com/content/book/9780071844611/chapter/chapter8

NICC. (2022). *Wildfires and Acres.* National Interagency Fire Center. https://www.nifc.gov/fire-information/statistics/wildfires

NIFC. (2022). *Suppression Costs | National Interagency Fire Center.* National Interagency Fire Center. https://www.nifc.gov/fire-information/statistics/suppression-costs

Perry, G. L. W. (1998). Current approaches to modelling the spread of wildland fire: A review. *Progress in Physical Geography: Earth and Environment*, *22*(2), 222–245. https://doi.org/10.1177/030913339802200204

Smith, A. B. (2020). *U.S. Billion-dollar Weather and Climate Disasters, 1980—Present (NCEI Accession 0209268)* [Data set]. NOAA National Centers for Environmental Information. https://doi.org/10.25921/STKW-7W73

The Matplotlib development team. (2023). *Matplotlib.animation.FuncAnimation.* Matplotlib 3.6.3 Documentation. https://matplotlib.org/stable/api/_as_-gen/matplotlib.animation.FuncAnimation.html

Verzoni, A. (2021). *Global Wildfire Deaths.* http://www.nfpa.org/News-and-Research/Publications-and-media/NFPA-Journal/2021/Winter-2021/News-and-Analysis/Dispatches/International

Wikipedia Contributors. (2022). Moore neighborhood. In *Wikipedia*.
https://en.wikipedia.org/w/index.php?title=Moore_-
neighborhood&oldid=1128664897

Zant, P. V. (2014). Doping. In *Microchip Fabrication* (6th ed.). McGraw-Hill
Education.
https://www.accessengineeringlibrary.com/content/book/9780071821018/
chapter/chapter11

A. Python Programming Environments

As discussed in Chapter 1, there are several ways to establish a good programming environment that uses Python:

1. We can download and install a Python distribution, such as the Anaconda Distribution and then use the IPython command line shell that comes with that distribution. This allows the user to execute Python interactively, one code line at a time.
2. Install a Python distribution, such as Anaconda, that comes with a Jupyter server then use Jupyter notebooks to compose executable code and text blocks.
3. Install a Python distribution, such as Anaconda, then use a plain text editor (notepad, gedit, TextEdit) application to compose code and then execute the code from the command line using the appropriate command line console for your operating system.
4. Install a Python distribution, then install an IDE application (PyCharm, Spyder, VSCode) that bundle a smart text editor, an interactive console, and enhanced debugging tools into a nice GUI.
5. Use a cloud computing platform such as CoCalc, pythonanywhere, or Google Colab that allows the user to use Jupyter notebooks without installing any software, except for a browser. The cloud service maintains the Python distribution.

The last method was discussed in detail in Chapter 1. We will consider the other methods here.

A.1. Installing a Python Distribution

There are several Python distributions used for scientific and engineering application development. The most common ones include:

- Anaconda (Anaconda, Inc., 2024)
- Enthought (Enthought, Inc, 2024)
- Active State (ActiveState Software Inc., 2024)

We will focus on the Anaconda distribution here. Setting up a working programming environment based on the Anaconda distribution should be straightforward.

1. Navigate to the Anaconda download page.
2. Fill in your email information and submit.
3. Choose the download for your operating system.
4. After the installer downloads, open it up and follow the instructions. Accept the default locations for files.

Certainly, one major benefit of programming in Python is the large number of packages available. These packages, such as matplotlib, allow for developing powerful scientific applications, but they are not all preinstalled when you perform the default installation of your distribution. You will need to install the packages to be used by this textbook.

Each package will have a set of dependencies (other required packages). To ensure that these dependencies are also installed a package manager app is used. Anaconda comes with a package manager called conda. Instead of using conda directly, we will use a conda graphical user interface app called Navigator, which is installed with the Anaconda distribution. Navigator will help you to manage the installed packages without learning all the details of the conda app.

Locate the Navigator app that was installed with the Anaconda distribution and start it up. Figure A.1 shows the Navigator home screen. The rectangular application tiles are additional apps that can be installed using Navigator. We will ignore these apps for now.

Now click on the Environments tab. Some Python packages have inconsistent dependencies. To avoid this problem, the Anaconda distribution allows separate programming environments to be set up depending on the project you are work-

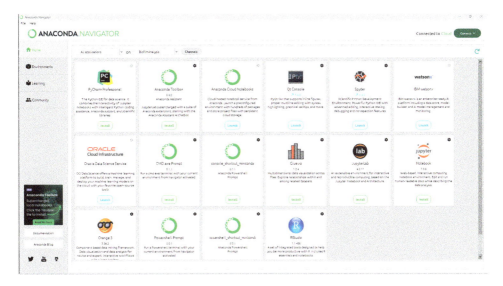

Figure A.1.: Navigator home screen.

ing on. These programming environments exist in separate file structures on your computer and are isolated from each other. Figure A.2 shows the default Environments screen. By default, there is one environment set up labelled "base (root)". While we can add packages to this environment, if we want, a better practice is to create a separate environment for each major programming project. We will create an environment that should work for all the examples and problems in the textbook.

To create a new environment,

1. Click on the "Create" tool at the bottom of the Environments screen.
2. Type in an appropriate name, such as "CompSciBook", and select the version of Python you wish to use. The default Python version should be acceptable for now.
3. Select "Not installed" from the package filter drop down menu at the top.
4. Search for the matplotlib, which is the first package listed in Table A1-1.
5. Select the package
6. Click on "Apply".

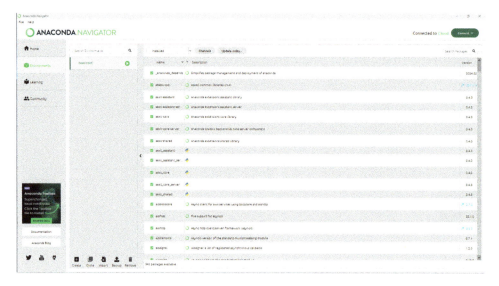

Figure A.2.: Default Environment screen.

Before repeating this process for the other packages, you should add an additional channel for Navigator to use.

1. Click on Channels
2. Type in conda-forge
3. Click on Update channels

Now repeat the above process for adding packages for the other packages that are not already installed. You will find that installing matplotlib will automatically install numpy, since it is a dependency.

There are several ways to execute Python code using the created environment. One method is to start Navigator, go to the Environments page, and select the green arrow, pointed out in Figure A.3.

Clicking the arrow tool gives you access to several choices for running Python. The Terminal gives you access to the operating system command prompt, which will allow you to execute a Python program from a file. The IPython selection opens up

Table A.1.: Required packages for textbook problems.

Package
matplotlib
numpy
pandas
scipy
seaborn
statsmodels
numdifftools

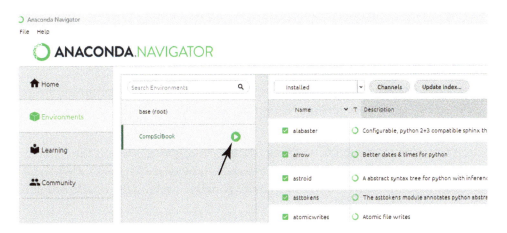

Figure A.3.: The execute Python button is shown by the arrow.

the IPython interactive interpreter from which you can execute individual Python commands.

A.2. Using Jupyter Notebooks with The Anaconda Distribution

Jupyter Notebook is an application for interactive code development and commentary. The computational notebook format is a file that has cells for executable code and output and cells for text that can be formatted using the Markdown language. The notebooks open in a browser where the executable code is run on a cloud-based server or alternatively on your computer, if the appropriate Jupyter Notebook application has been installed.

When the Anaconda distribution is installed on your computer, the Jupyter Notebook application should be installed by default and will execute code from a browser using the base environment. One way to create a Jupyter Notebook file is to

1. start Navigator
2. go to the Environments window
3. click on the base environment
4. click on the green arrow for the environment drop-down menu
5. select "Open with Juptyer Notebook", as shown in Figure A.4.

To create a Jupyter notebook that uses another environment, such as the CompSci-Book environment, use Navigator to install the Jupyter package into the selected environment. You can now start a Jupyter Notebook session from selected environment tab in Navigator.

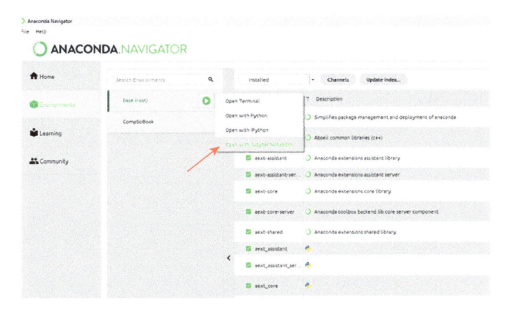

Figure A.4.: The Navigator environments window. The arrow points out the selection to start a Jupyter notebook using the base environment.

A.3. Use the Spyder IDE with the Anaconda Distribution

As the complexity of the coding project increases, there will likely come a time when it is advantageous to shift the programming environment away from computational notebooks (Jupyter Notebook, Google Colab notebooks, etc.) towards a more robust integrated development environment (IDE) that includes a code smart text editor, an interactive test run environment, and enhanced debugging tools. One such IDE that comes with the Anaconda distribution is the Spyder app that will be installed automatically in the base environment. The Spyder documentation offers a good overview of the basic features available (Spyder Doc Contributors, 2024). Figure A.5 shows the default opening screen.

The Spyder text editor will color Python key words and variable names that have been used. It will check syntax for errors.

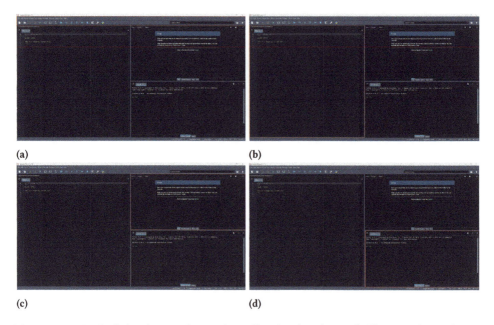

(a) (b)

(c) (d)

Figure A.5.: Default Spyder windows. a) Toolbar (outlined in red); b) Text Editor; c) Help Window; d) IPython console.

One method of using Spyder with a different environment is to use Navigator to install the Spyder package in the appropriate environment. A link to the specific version of Spyder will be added n the application listing. You can also change the environment used by Spyder in the Preferences tab.

A.4. References

ActiveState Software Inc. (2024). ActiveState Python Distribtution. *ActiveState.* https://www.activestate.com/products/python/

Anaconda, Inc. (2024). *Anaconda Distribution.* Anaconda. https://www.anaconda.com/download

Enthought, Inc. (2024). *Enthought Downloads.*

https://assets.enthought.com/downloads/

Spyder Doc Contributors. (2024). *Quickstart—Spyder 5 documentation.* https://docs.spyder-ide.org/current/quickstart.html

B. Advanced Plotting with Matplotlib

In this appendix, we will explore the following advanced visualization topics:

- the matplotlib object-oriented interface
- creating subplots
- using logarithmic scaled axes
- adding error bars to data points
- adjusting text properties using TeX

B.1. The Two Matplotlib Interfaces

Up until this point we have used the Matplotlib interface functions available in the pyplot submodule. This interface was designed to duplicate the user interface in the application MATLAB, a proprietary programming language and programming environment used by many engineers and scientists. The pyplot interface uses a procedural programming model where a program executes using the control structures we introduced earlier: sequential, conditional, and iterative, and where blocks of code that can be reused are packaged in a procedure, which is a function in Python.

The other programming model that can be used with Matplotlib is the Object-Oriented Programming model (OOP). In the OOP model a program is composed of objects that have properties and codes (called methods) that can change properties of the object. At a fundamental level, Python is an object-oriented programming language, although we have not been emphasizing that aspect until now. Even basic data types such integers, floats, and strings are objects in Python.

Two important objects that are used to create matplotlib graphs are the Figure object and the Axes object. An instance of the Figure object can contain one or more Axes objects. The Axes object is a rectangular area that will hold the elements of a graph: x-axis line, y-axis line, data symbols, lines, etc. We can create an empty Figure object called fig with the code

```
fig = plt.figure()
```

Note that we can use any legal variable name we want for the Figure object. Calling it fig is a common choice.

An Axes object can be created using the add_subplot method of the Figure object. Here is an example of creating an Axes object named ax1 in the Figure instance fig.

```
ax1 = fig.add_subplot(1, 1, 1)
```

A plot can be added to an Axes object using the plot method for Axes objects. Let's put this all together and create a simple graph using the Object-Oriented interface. The code is in Figure B.5 and the resulting figure is shown as Figure B.2.

The next section will describe details of using the add_subplot method.

B.2. Subplots

There will be many times in communicating scientific information that two or more different properties can best be compared by putting them in the same figure. If both properties are to be visualized by a 2d graph then we need to place multiple graphs in the same figure, so that comparisons can be easily made. Let's consider a biomechanics-related example. Suppose we want to understand the mechanics of performing a standing long jump. A simple first step might be to video a person performing the motion and extract some of the quantitative motion data from the video. Figure B.3 shows a man performing a standing long jump. Using video analysis software, a researcher obtained the position as a function of time of the jumper's approximate center of mass. One interesting comparison to make is the y-component of the position and the velocity as functions of time. Since these are physically different properties they need to be on different graphs, but to aid in

```
1  """
2  Program: Simple Example of Subplot
3  Author: C.D. Wentworth
4  Version: 2.20.2022.1
5  Summary:
6      This program illustrates the basic syntax for creating a Figure object
7      instance and an Axes object instance.
8  Version History:
9      2.20.2022.1: base
10 """
11 import matplotlib.pyplot as plt
12 import numpy as np
13
14 # Set up numpy random number generator
15 np_rng = np.random.default_rng(seed=314159)
16
17 # Create some random data
18 randomData = np_rng.random(100)
19
20 # Plot the data
21 fig = plt.figure()
22 ax1 = fig.add_subplot(1,1,1)
23 ax1.plot(randomData)
24 plt.savefig('randomPlot.png' , dpi=300)
```

Figure B.1.: Code for setting up subplots using the OOP interface.

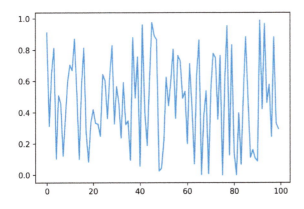

Figure B.2.: Output from the code of Figure B.1.

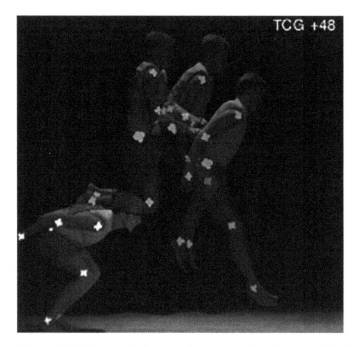

Figure B.3.: Composite image of a man performing a standing long jump.

comparing them, we should put the two graphs in the same figure, as shown in Figure B.4.

How was this figure created? We are going to use the Objected-Oriented Matplotlib interface discussed above. Figure B.5 shows a listing of the program that created B.4. In line 20, we create an instance of the figure class that is defined in the matplotlib library, and we call the instance fig. All the methods associated with the figure class can now be used by the fig instance. We will use some of those methods to create the desired figure.

In line 21 we use the add_subplot() method to create a subplot in fig, and we call the subplot f1. The meaning of the arguments are given in the following

```
add_subplot(nrows, ncols, index, **kwargs)
```

where

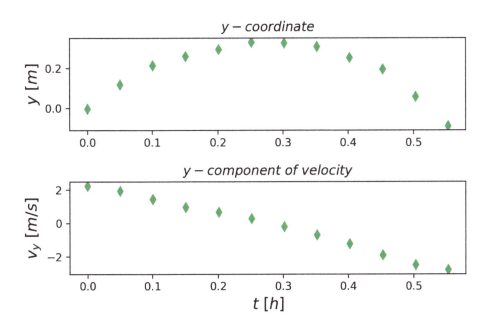

Figure B.4.: y-component of position and velocity for the center of mass of a standing long jumper.

```python
1  """
2  Title: Standing Long Jump Subplots
3  Author: C.D. Wentworth
4  Version: 03.1.2019.1
5  Summary: This program illustrates a method for creating
6          subplots in a single figure.
7  """
8  import numpy as np
9  import matplotlib.pylab as plt
10
11 #--Main Program
12
13 # Read in data
14 data = np.loadtxt("StandingLJ_Data.txt",skiprows=7)
15 tData = data[:,0]
16 yData = data[:,2]
17 vyData = data[:,4]
18
19 # Plot data
20 fig = plt.figure()
21 f1 = fig.add_subplot(2,1,1)
22 f1.plot(tData,yData,linestyle='',marker='d',
23 color='g',markersize=6.0)
24 #f1.set_xlabel(r'$t [h]$',fontsize=14)
25 f1.set_ylabel(r'$y \ [m]$',fontsize=14)
26 f1.set_title(r'$y-coordinate$')
27 f2 = fig.add_subplot(2,1,2)
28 f2.plot(tData,vyData,linestyle='',marker='d',
29             color='g',markersize=6.0)
30 f2.set_xlabel(r'$t \ [h]$',fontsize=14)
31 f2.set_ylabel(r'$v_y \ [m/s]$',fontsize=14)
32 f2.set_title(r'$y-component \ of \ velocity$')
33 plt.tight_layout()
34 plt.savefig('StandingLJ_subplots.png',dpi=300)
35 plt.show()
```

Figure B.5.: Code for the standing long jump subplots.

nrows, the number of rows in the figure

ncols, the number of columns in the figure

index, a positive integer indicating the subplot being constructed

**kwargs, corresponds to a variety of optional keyword arguments

So, add_subplot(2,1,1) means that we are creating a figure with 2 rows and 1 column for subplots, and we are currently focused on the first subplot, which would be the top one. Line 22-23 creates the actual subplot. Lines 25 and 26 illustrate how we can add properties to the subplot. Line 27 creates the second subplot by using the index 2 as the third argument value. The set_tight_layout(True) method applied in line 33 allows padding to be added between parts of a figure. It has other possible arguments that can be specified to give the programming considerable control over the appearance, but often the value True will achieve the desired results.

The xlabel and ylabel arguments in the code of Figure B.5 illustrate how to use the mathematics typesetting language TeX to create labels that contain more sophisticated mathematical formatting. A string enclosed by $ signs, as in '$v_y \ [m/s]$', is interpreted as TeX code. The matplotlib documentation gives some basic examples of using TeX to create professional mathematical formatting (The Matplotlib Development Team, 2023b).

B.3. Uncertainty Bars

Quantitative experimental measurements always carry some uncertainty due to random variations that can occur when the same measurement is repeated. If a measurement is repeated several times to give a sample of the measurement, then typically the sample mean will be used as the value associated with the data point and the uncertainty in the mean due to random variation is given by the sample standard error. In a graph of the measurements, we want to communicate the estimated uncertainty in a particular data point, and we can do this by adding an uncertainty, or error, bar above and below the actual data point. This communicates to the reader that the actual value has a range of possible values.

Table B.1.: Position-time data.

t[s]	x [m]	Dx [m]
0	0	0
0.2	0.094	0.06
0.4	0.336	0.25
0.6	0.46	0.4
0.8	1.344	0.2
1	2.1	0.42
1.2	3.324	0.32
1.4	4.116	0.32
1.6	5.376	0.42
1.8	6.804	0.42
2	8.4	0.42

As a specific example, consider vertical position data taken from a video of a dropped ball Table B.1 gives the data from the data file.

Figure B.6 shows a plot of the ball's position as a function of time. The uncertainty of each position measurement is indicated with the red bar drawn through the data points. The value of the uncertainty was read from the data file in the third column.

The key to adding error bars is to use the errorbar() method instead of the plot method when constructing the plot. Most of the arguments of errorbar() are the same as for plot(), but there is the additional keyword argument yerr, which specifies the list containing the uncertainty estimates for each data point. Line 21 of Figure B.7 shows how errorbar() was used to create the graph in Figure B.6.

The matplotlib documentation gives many examples of other plot types that might be useful (The Matplotlib Development Team, 2023a).

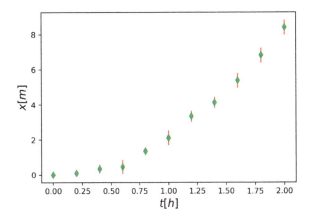

Figure B.6.: Position as a function of time for a ball dropped from rest vertically.

```python
"""
Title: Plot Data Error Bars
Author: C.D. Wentworth
Version: 12.31.2018.1
Summary: This program illustrates how to add errobars
         (or uncertainty bars) to a data point based on
         uncertainty estimates imported from the data file.

"""
import numpy as np
import matplotlib.pyplot as plt
# Read in data
data = np.loadtxt("droppedBallData.txt",skiprows=2)
tData = data[:,0]
xData = data[:,1]
dxData = data[:,2]

# Plot data
fig = plt.figure()
f1 = fig.add_subplot(1,1,1)
f1.errorbar(tData,xData,linestyle='',marker='d',
        color='g',markersize=6.0,yerr=dxData, ecolor='r')
f1.set_xlabel(r'$t [h]$',fontsize=14)
f1.set_ylabel(r'$x [m]$',fontsize=14)
plt.savefig('droppedBallWithErrorBars.png',dpi=300)
```

Figure B.7.: Code for creating error bars from droppedBallPlotWithErrorBars.py.

B.4. References

The Matplotlib Development Team. (2023a). *Plot types—Matplotlib 3.7.1 documentation.* https://matplotlib.org/stable/plot_types/index.html

The Matplotlib Development Team. (2023b). *Writing mathematical expressions—Matplotlib 3.7.1 documentation.* https://matplotlib.org/stable/tutorials/text/mathtext.html

Index

Index

About the Author

Christopher D. Wentworth is Professor Emeritus of Physics at Doane University, where he taught physics, computational science, and engineering courses for over 34 years. He earned a Bachelor of Science in Physics from Duke University and a Ph.D. in theoretical physics from Florida State University. His research interests while in graduate school and during post-doctoral work at Florida State University and at the University of Leiden involved developing models of magnetic materials using the tools of quantum statistical mechanics. This research was heavily computational in nature involving computational symbolic mathematics and Monte Carlo simulations. After moving to Doane University (then called Doane College), his major interest became physics education research. Developing interest and expertise in computational science was an important part of this work and continues to be a primary teaching interest. More recently, his research includes developing mathematical and computational models of biofilm growth and other topics in condensed matter physics and biological physics.

www.ingramcontent.com/pod-product-compliance
Lightning Source LLC
LaVergne TN
LVHW061954050326
832903LV00036B/4827